2017年"一流应用技术大学"建设系列教材

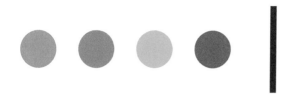

机械加工、成型和安装工艺

Machining, Forming and Assembly Technology

主　编　李　妍　薛　静　李龙泉
副主编　张　凯　杨筱晶

西安电子科技大学出版社

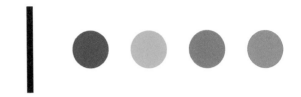

内容简介

机械加工工艺、成型工艺和安装工艺是机械制造及其相关专业的专业基础课。本书参考了大量机械制造行业的有关规范，依据机械制造专业教学和生产特点，引进西班牙机床学院(IMH)课程标准，充分利用引导式教学方法促进学生对专业基础知识的理解和掌握。本书突出专业知识的广泛性，全面讲解材料的种类、特点、热处理的工序和工艺，铸造、锻造、冲压等常见成型加工的分类、设备类型，车削、铣削、磨削、特种加工等加工工艺的分类、设备类型，机械加工工艺规程，装配工艺基础与车间管理等知识。

本书可作为职业学校机械类专业的教材，也可作为在职培训或自学用书。

图书在版编目(CIP)数据

机械加工、成型和安装工艺 / 李妍，薛静，李龙泉主编. -- 西安：西安电子科技大学出版社，2024.1
ISBN 978-7-5606-5191-0

Ⅰ. ①机… Ⅱ. ①李… ②薛… ③李… Ⅲ. ①金属切削 Ⅳ. ①TG506

中国版本图书馆CIP数据核字(2018)第278376号

策划编辑　毛红兵　万晶晶
责任编辑　权列秀　万晶晶
出版发行　西安电子科技大学出版社(西安市太白南路2号)
电　　话　(029)88242885　88201467　　　邮　编　710071
网　　址　www.xduph.com　　　电子邮箱　xdupfxb001@163.com
经　　销　新华书店
印刷单位　广东虎彩云印刷有限公司
版　　次　2024年1月第1版　　2024年1月第1次印刷
开　　本　787毫米×1092毫米　1/16　印张　14
字　　数　332千字
定　　价　54.00元
ISBN　978-7-5606-5191-0 / TG
XDUP　5493001-1
***** 如有印装问题可调换 *****

前言

本书以机械制造类型为导向依据，引进西班牙机床学院(IMH)课程标准，充分利用引导式教学方法促进学生对专业基础知识的理解和掌握。本书采用中英文混合编写，对关键词句及重要知识点进行英语翻译标注，在加强专业基础知识学习的同时，注重能力和专业素质的培养，使本书更加适用于国际化专业人才培训。

本书突出专业知识的广泛性，全面讲解材料的种类、特点、热处理的工序和工艺，铸造、锻造、冲压等常见成型加工的分类、设备类型，车削、铣削、磨削、特种加工等加工工艺的分类、设备类型，机械加工工艺规程，装配工艺基础与车间管理等知识。

本书注重培养学生对机械专业知识的认知与理解，包含并融入行业岗位所需的相关理论知识，是一本以机械加工、成型加工为主的理论课教材。

本书是天2017年"一流应用技术大学"建设系列教材，由李妍、薛静、李龙泉担任主编，张凯、杨筱晶担任副主编。编写分工为：薛静编写第一章；李妍编写第二章、第三章、第四章、第五章、第七章、第十章及第十二章；张凯编写第六章；杨筱晶编写第八章；李龙泉编写第九章、第十一章。

由于编者理论水平和教学经验有限，书中难免有不妥之处，衷心希望同行或读者不吝赐教，及时批评指正。

编者
2020.11

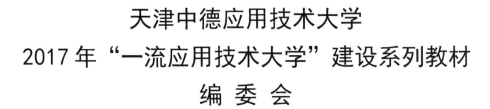

天津中德应用技术大学
2017年"一流应用技术大学"建设系列教材
编 委 会

主　任：徐琤颖

委　员：（按姓氏笔画排序）

　　　　王庆桦　王守志　王金凤　邓　蓓　李　文

　　　　李晓锋　杨中力　张春明　陈　宽　赵相宾

　　　　姚　吉　徐红岩　靳鹤琳　薛　静

目录

第一篇 工程材料 / 1

 第一章 金属材料 / 3

 第二章 铁碳合金与钢的热处理 / 19

第二篇 成型加工与压力加工 / 27

 第三章 铸造 / 29

 第四章 锻造 / 48

 第五章 冲压 / 67

第三篇 切削加工与特种加工 / 79

 第六章 车削加工 / 81

 第七章 铣削加工 / 113

 第八章 磨削加工 / 141

 第九章 特种加工 / 155

第四篇 机械加工工艺规程与车间管理 / 169

 第十章 机械加工过程与工艺规程 / 171

 第十一章 装配工艺基础 / 187

 第十二章 车间布局与 5S 管理 / 193

参考文献 / 218

第一篇　工程材料

人类早在 6000 年以前就掌握了金属冶炼的方法。古埃及人在公元前 4000 年便掌握了炼铜技术。我国在夏代早期也掌握了青铜冶炼技术，在春秋战国时期，已经开始大量使用铁器。

本篇介绍金属材料的特性及其热处理。

第一章 金属材料

1.1 概述

工程材料有各种不同的分类方法，一般都将工程材料按化学成分进行分类。

非金属材料也是重要的工程材料。它包括耐火材料、耐火隔热材料、耐蚀（酸）非金属材料和陶瓷材料等。

金属材料是最重要的工程材料，包括金属和以金属为基的合金。工业上把金属和其合金分为两大类：

(1) 黑色金属材料：铁和以铁为基的合金（钢、铸铁和铁合金）。

(2) 有色金属材料：黑色金属以外的所有金属及其合金。

应用最广的是黑色金属。以铁为基的合金材料占整个结构材料和工具材料的 90% 以上。黑色金属材料的工程性能比较优越，价格也比较便宜，是最重要的工程金属材料。

有色金属按照性能和特点可分为轻金属、易熔金属、难熔金属、贵金属、稀土金属和碱土金属，它们是重要的有特殊用途的材料。

复合材料就是用两种或两种以上不同材料组合的材料，其性能是其他单质材料所不具备的。它在强度、刚度和耐蚀性方面比单纯的金属、陶瓷和聚合物都优越，是特殊的工程材料，具有广阔的发展前景。

为有机合成材料，也称聚合物。它具有较高的强度、良好的塑性、较强的耐腐蚀性能、很好的绝缘性和重量轻等优良性能，是发展最快的一类新型结构材料。高分子材料种类很多，工程上通常根据机械性能和使用状态将其分为三大类：塑料、橡胶和合成纤维。

Metal materials are the most important engineering materials, including metals and metal-based alloys. In industry, metals and their alloys are divided into two main categories:

1. Ferrous metal materials: iron and iron-based alloys (steel, cast iron and ferroalloys).
2. Non-ferrous metal materials: all metals except ferrous metals and their alloys.

The most widely used are ferrous metals. Iron-based alloys account for more than 90% of all structural and tool materials. Ferrous metal material are the most important engineering metal materials because of their superior engineering performance and low price.

Non-ferrous metals can be divided into light metals, fusible metals, refractory metals, precious metals, rare earth metals and alkaline earth metals according to their properties and characteristics. They are important materials for special purposes.

Non-metallic materials are also important engineering materials. They include refractories, refractories and heat insulation materials, corrosion resistant (acid) non-metallic materials and ceramic materials.

Composite materials are composed of two or more different materials, whose properties are not available in other materials. Composites can be made up of a variety of materials. It is superior to pure metals, ceramics and polymers in strength, stiffness and corrosion resistance. It is a special engineering material and has broad prospects for development.

Macromolecule materials are organic synthetic materials, also known as polymers. It has high strength, good plasticity, strong corrosion resistance, good insulation and light weight and other excellent properties. It is a new type of structural materials with the fastest development in engineering. There are many kinds of macromolecule materials, which are usually classified into three major categories according to their mechanical properties and service conditions: plastics, rubber and synthetic fibers.

1.2　金属材料的性能

为什么在交通运输、冶金设备、石油化工设备、农用机械、各种机械加工设备、动力设备中，金属制品占80%～90%，成为现代制造业中最主要的材料呢？

金属材料之所以获得如此广泛的应用，主要是由于它具有制造机器所需要的物理、化学和力学性能，并且可用较简便的工艺方法加工成适用的机械零件，亦即具有所需的工艺性能。

机械制造中所用的金属材料以合金为主，很少使用纯金属，原因是合金比纯金属具有更好的力学性能和工艺性能，且价格低廉。合金是以一种金属为基础，加入其他金属或非金属，经过熔炼或烧结制成的具有金属特性的材料。最常用的合金是以铁为基础的铁碳合金，如碳素钢、合金钢、灰铸铁等；还有以铜或铝为基础的黄铜、青铜、硅铝明等。

机械加工、成型和安装工艺

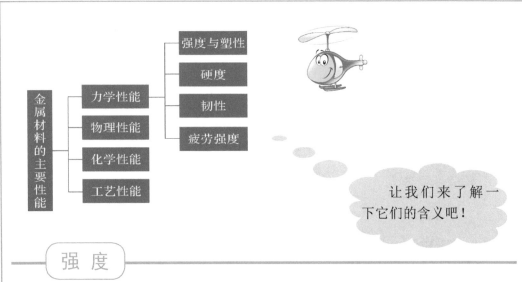

强 度

强度是指金属材料在力的作用下，抵抗塑性变形和断裂的能力。

按外力作用方式的不同，强度可分为抗拉强度、抗压强度、抗弯强度、抗剪强度和抗扭强度等，其中以抗拉强度最为常用。当材料承受拉力时，强度主要是指屈服强度和抗拉强度。

(1) 屈服强度：拉伸试样产生屈服现象时的应力，用符号 σ_s 表示。

(2) 抗拉强度：金属材料在拉断前所能承受的最大应力，用符号 σ_b 表示。

屈服强度和抗拉强度在选择、评定金属材料及设计机械零件时具有重要意义。由于机器零件或构件工作时，通常不允许发生塑性变形，因此多以屈服强度作为强度设计的依据。对于脆性材料，因断裂前基本不发生塑性变形，故无屈服强度可言，在强度计算时，则以抗拉强度为依据。

塑 性

塑性是指金属材料产生塑性变形而不被破坏的能力，通常以伸长率 δ 来表示。

评定金属材料塑性好坏，除伸长率外，还可用断面收缩率（符号：ψ）表示。

伸长率和断面收缩率值愈大，材料的塑性愈好。良好的塑性不仅是金属材料进行轧制、锻造、冲压、焊接的必要条件，而且在使用时万一超载，由于产生塑性变形，能够避免断裂。

硬 度

硬度是衡量金属材料软硬程度的一项重要的性能指标，它既可以理解为材料抵抗弹性变形、塑性变形或破坏的能力，也可以表述为材料抵抗残余变形和反破坏的能力。硬度是材料弹性、塑性、强度和韧性等力学性能的综合指标。

金属硬度 (Hardness) 的代号为 H。按硬度试验方法的不同，硬度的常规表示有布氏硬度(HB)、洛氏硬度(HR)、维氏硬度(HV)等，其中以HB及HR较为常用。

布氏硬度(HB)一般用于表示硬度较低的材料，如有色金属、热处理之前或退火后的钢铁。洛氏硬度(HR)一般用于表示硬度较高的材料，如热处理后的硬度等。

让我们通过动画来了解一下吧！

布氏硬度

布式硬度(HB)是以一定大小的试验载荷，将一定直径的淬硬钢球或硬质合金球压入被测金属表面，保持规定时间，然后卸荷，测量被测表面压痕直径来测定材料硬度值的。

洛氏硬度

洛式硬度是以压痕塑性变形深度来确定硬度值的。

根据试验材料硬度的不同，有三种不同的标度：

HRA：采用60 kg载荷和钻石锥压入器测得的硬度，用于硬度较高的材料（如硬质合金等）。

HRB：采用100 kg载荷和直径1.58 mm淬硬的钢球测得的硬度，用于硬度较低的材料（如退火钢、铸铁等）。

HRC：采用150 kg载荷和钻石锥压入器测得的硬度，用于硬度很高的材料（如淬火钢等）。

韧 性

韧性是指金属材料在塑性变形和破裂过程中吸收能量的能力。金属韧性可以分为冲击韧性和断裂韧性。

疲劳强度

疲劳强度是指材料在无限多次交变载荷作用下不会产生破坏的最大应力，也称为疲劳极限。实际上，金属材料并不可能作无限多次交变载荷试验。

机械上的许多零件，如曲轴、齿轮、连杆、弹簧等是在周期性或非周期性动载荷（称为疲劳载荷）的作用下工作的。这些承受疲劳载荷的零件发生断裂时，其应力往往大大低于该材料的强度极限，这种断裂称作疲劳断裂。

工艺性能

工艺性能是金属材料物理、化学性能和力学性能在加工过程中的综合反映，是指是否易于进行冷、热加工的性能。按工艺方法的不同，工艺性能可分为铸造性能、锻造性能、焊接性能和切削加工性能等。

物理性能

物理性能是指金属材料不发生化学反应就能表现出来的一些性能，如密度、熔点、导电性、导热性、磁性和热膨胀性等。

化学性能

化学性能是指金属材料在室温或高温时抵抗各种介质的化学侵蚀能力，主要有耐腐蚀性、抗氧化性和化学稳定性。

Strength

Strength is the ability of metal materials to resist plastic deformation and fracture under the action of force.

According to the action mode of external force, it can be divided into tensile strength, compressive strength, flexural strength, shear strength and torsion strength, among which tensile strength is the most commonly used. When the material bears tension, strength refers mainly to yield strength and tensile strength.

(1) Yield strength: refers to the stress of the tensile specimen when yielding occurs. Symbol: σ_s

(2) Tensile strength: refers to the maximum stress that a metal material can bear before it breaks. Symbol: σ_b

Yield strength and tensile strength are of great significance in selecting and evaluating metal materials and designing mechanical parts. Because plastic deformation is not usually allowed when machine parts or components are working, yield strength is often used as the basis for strength design. For brittle materials, no plastic deformation occurs before fracture, so there is no yield strength. When calculating strength, it is based on tensile strength.

Plasticity

It refers to the ability of metal materials to produce plastic deformation without being destroyed. It is usually expressed as elongation δ.

In addition to elongation, section shrinkage(symbol: ψ) can also be used to evaluate the plasticity of metal materials.

The greater the elongation and shrinkage, the better the plasticity of the material. Good plasticity is not only a necessary condition for rolling, forging, stamping and welding metal materials, but also can avoid fracture in case of overload in use due to plastic deformation.

Hardness

Hardness is an important performance indicator to measure the degree of hardness and softness of metal materials. It can be understood as the ability of materials to resist elastic deformation, plastic deformation or failure, and also expressed as the ability of materials to resist residual deformation and anti-failure.

Hardness is a comprehensive indicator of mechanical properties such as elasticity, plasticity, strength and toughness.

Hardness is coded H. According to the different hardness test methods, the conventional expressions are Brinell hardness (HB), Rockwell hardness (HR), Vickers hardness (HV), among which HB and HR are commonly used.

Brinell hardness (HB) is generally used when materials are soft, such as nonferrous metals, steel before or after heat treatment. Rockwell hardness (HR) is generally used for materials with higher hardness, such as hardness after heat treatment and so on.

Brinell hardness

Cloth hardness (HB) is a test load of a certain size. Hardened steel balls or cemented carbide balls of a certain diameter are pressed into the surface of the metal to be measured. The required time is maintained, then unloaded, and the indentation diameter of the surface to be measured is measured.

Rockwell hardness

Rockwell hardness is determined by indentation plastic deformation depth. According to the different hardness of the test material, it can be expressed by three different scales:

HRA: The hardness obtained by 60 kg load and diamond cone indenter is used for materials with very high hardness (such as cemented carbide).

HRB: The hardness obtained by hardening steel balls with a diameter of 1.58 mm and a load of 100 kg is used for materials with lower hardness (such as annealed steel, cast iron, etc.).

HRC: The hardness obtained by 150 kg load and diamond cone indenter is used for materials with high hardness (such as quenched steel).

Toughness

Toughness refers to the ability of metal materials to absorb energy in the process of plastic deformation and fracture. Metal toughness can be divided into impact toughness and fracture toughness.

Fatigue strength

Fatigue strength refers to the maximum stress of a material under an infinite number of alternating loads without failure, which is called fatigue strength or fatigue limit. In fact, it is impossible for metal materials to be subjected to an infinite number of alternating load tests. Many mechanical parts, such as crankshaft, gear, connecting rod and spring, work under periodic or non-periodic dynamic loads (called fatigue loads). When these parts under fatigue load break, their stress is often much lower than the strength limit of the material, which is called fatigue fracture.

Processing property

Processing properties are the comprehensive reflection of physical, chemical and mechanical properties of metal materials in the process of processing. They refer to whether it is easy to cold and hot working. According to the different processing methods, it can be divided into casting performance, forgeability, welding performance and cutting performance.

Physical property

Physical properties refer to some properties that metal materials can show without chemical reaction, such as density, melting point, conductivity, thermal conductivity, magnetism and thermal expansion.

Chemical properties

Chemical properties refer to the ability of metal materials to resist chemical erosion of various media at room or high temperature, mainly including corrosion resistance, oxidation resistance and chemical stability.

1.3　金属的晶体结构

按物质内部原子的聚集状态的不同，固态物质分为晶体和非晶体。

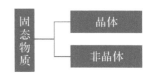

晶体与非晶体有什么区别呢？

晶体：内部原子在空间按一定规则排列的物质，如金刚石、石墨及固态的金属、合金等。晶体具有固定的熔点及各向异性等特征。例如：铁的熔点为1538℃，铜的熔点为1083℃。

非晶体：内部原子排列无一定规则的物质，如玻璃、石蜡、松香等。非晶体物质没有固定的熔点，而且性能无方向性。

这就是晶体吗？

是的，让我们先来了解一下液态金属的结晶过程吧！

晶体的性能随着原子的排列方位而改变，即**晶体具有各向异性**。晶体不同方向上性能不同的性质叫做**晶体的各向异性**。

液态金属结晶过程是遵循"晶核不断形成和长大"这个结晶基本规律进行的。

开始时，液体中先出现一些极小晶体，我们称其为**晶核**。在这些晶核里，有些是依靠原子自发地聚集在一起的，它们按金属晶体固有规律排列而成，这些晶核称为**自发晶核**。金属的冷却速度愈快，自发晶核愈多。除此之外，液体中有些高熔点杂质形成的微小固体质点，也可起晶核作用，这种晶核叫做**外来晶核或非自发晶核**。

在晶核出现之后，液态金属的原子就以晶核为中心，按一定几何形状不断地排列起来形成**晶体**。晶体沿着各个方向生长，但是它的速度是不均匀的，通常按照一次晶轴、二次晶轴……呈树枝状长大。

在原有晶体长大的同时，在剩余液体中又陆续出现新的晶核，这些晶核也同样长大成晶体。这样就使液体愈来愈少。当晶体长大到与相邻的晶体互相抵触时，这个方向的长大便停止了。当全部晶体都彼此相遇、液体耗尽时，结晶过程结束。

好复杂的过程啊！！！

别担心！现在让我们以纯金属为例，再为大家做一下解释吧！

Amorphous: Material whose internal atoms are arranged irregularly, such as glass, paraffin, rosin, etc. Amorphous materials have no fixed melting point and no directional property.

Crystal: Material with regular arrangement of internal atoms in space, such as diamond, graphite, solid metal, alloy, etc. Crystals are characterized by fixed melting point and anisotropy. For example, the melting point of iron is 1538 degrees Celsius and that of copper is 1083 degrees Celsius.

Crystallization process of liquid metal: It follows the basic crystallization law of "continuous formation and growth of nuclei".

At the beginning, some very small crystals which first appeared in the liquid are called nuclei.

Some of these nuclei are spontaneously assembled by atoms, and they are arranged according to the inherent laws of metallic crystals. These nuclei are called spontaneous nucleation. The faster the metal is cooled, the more spontaneous nuclei are formed. In addition, the thiny solid particle formed by some high melting point impurities in the liquid can also play the role of nucleation, which is called foreign nucleation on non-spontaneous nucleation.

After the appearance of nuclei, the atoms of liquid metals take the crystal nuclei as the centre and are arranged in a certain peometry to form a crystal.

Crystals grow in all directions, but their velocities are not uniform, It usually grows in dendritic shape according to primary crystal axis, secondary crystal axis and so on.

At the original crystal grows, new nuclei appear in the remaining liquid, and these nuclei grow into crystals. This makes the liquid less and less. When the crystal grows up to conflict with the adjacent crystal, the growth in this direction stops. When all the crystals meet each other and the liquid is exhausted, the crystallization process ends.

1.4 纯金属的结晶过程

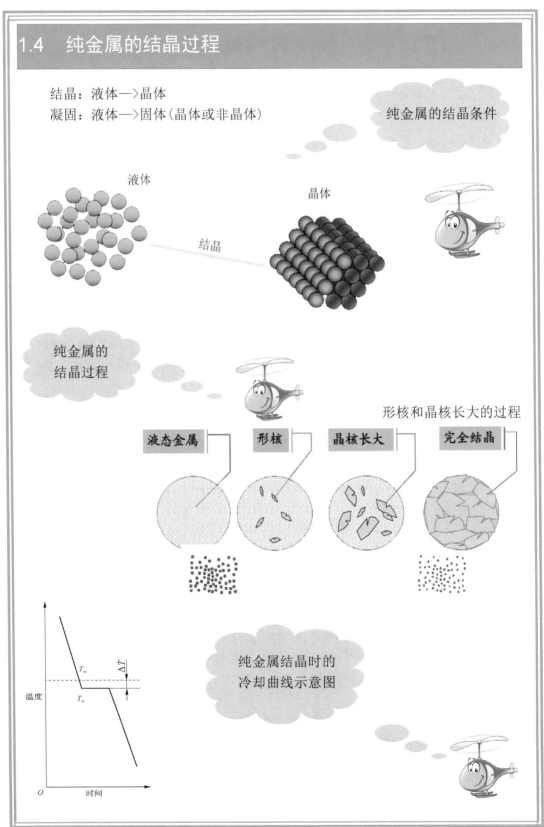

T_m：理论结晶温度

T_n：实际结晶温度

理论结晶温度与实际结晶温度之差称为过冷度。

过冷度的大小与冷却速度密切相关。冷却速度愈快，实际结晶温度就愈低，过冷度就愈大；反之，冷却速度愈慢，过冷度愈小。

过冷：实际结晶温度低于理论结晶温度（平衡结晶温度），这种现象称为过冷。

纯铁结晶过程示意图

 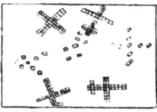 

 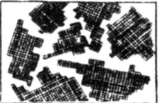 

下面让我们来了解一些关于晶体的基本概念吧！

简单立方晶格、晶胞示意图

晶 格

为了便于表明晶体内部原子排列的规律，把每个原子看成是固定不动的刚性小球，并用一些几何线条将晶格中各原子的中心连接起来，构成一个空间格架，各原子的中心就处在格架的几个结点上，这种抽象的、用于描述原子在晶体中排列形式的几何空间格架，简称晶格。

晶格

机械加工、成型和安装工艺

晶胞

由于晶体中原子有规则排列且有周期性的特点，为了便于讨论，通常只从晶格中选取一个能够完全反映晶体特征的、最小的几何单元来分析晶体中原子排列的规律，这个最小的几何单元称为晶胞。

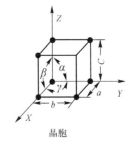
晶胞

晶轴

在晶体学中，通常取晶胞角上某一结点作为原点，沿其三条棱边作三个坐标轴 X、Y、Z，并称之为晶轴。

而规定坐标原点的前、后、上方为轴的正方向，反之为反方向，并以（晶格常数）棱边长度和棱面夹角表示晶胞的形状和大小。

整个晶格就是由许多大小、形状和位向相同的晶胞在空间重复堆积而成的。

 常见的晶格类型有哪些？

根据晶胞的三条棱边是否相等、三个夹角是否相等以及是否为直角，晶体学将所有晶体分为 7 个晶系，14 种空间点阵，称作布喇菲空间点阵。

常见的三种晶格类型为：体心立方晶格、面心立方晶格、密排六方晶格。

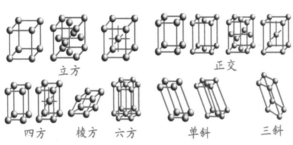

布喇菲空间点阵

立方　　正交

四方　棱方　六方　　单斜　　三斜

每一个晶核长成为一个晶体，这种晶体称为单晶体。

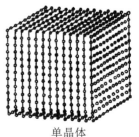

单晶体

在多晶体中，这些小颗粒晶体叫晶粒。
晶粒与晶粒之间的界面叫晶界。

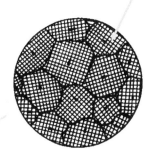

金属的晶粒大小对金属材料的力学性能、化学性能和物理性能影响很大。在一般情况下，晶粒越细小，则金属材料的强度和硬度越高，塑性和韧性越好。

通过这些介绍才知道，原来我们身边的金属材料的内部组织结构是这样的呀！

是不是长了很多知识呀！嘻嘻！

在我们的日常生活中，除了纯金属以外，还有哪些类型的金属材料为我们所用呢？

我们利用纯金属优良的导电性、导热性、化学稳定性及美丽的金属光泽等性能，制作出各种导电体、传热器、装饰品、艺术品等；但几乎各种金属的强度、硬度、耐磨性等力学性能都比较差，因而不适合制作对力学性能要求较高的机械零件和工模具。再加之纯金属的种类有限（大约只有79种），并且价格还高，因此在工业上应用较少。

实际上，现代工业中广泛应用的是合金，尤其是铁碳合金。

下一章我们将为大家介绍铁碳合金和钢的热处理。

Lattice

In order to show the regularity of the arrangement of atoms in crystals, each atom is regarded as a fixed rigid sphere, and the centers of atoms in the lattice are connected by some geometric lines to form a space lattice. The centers of atoms are located at several nodes of the lattice. This abstract description is used to describe atoms in the lattice. Geometric spatial lattices in crystal arrays are called lattices for short.

Unit cell

Because atoms in crystals are arranged regularly and periodically, for the sake of discussion, the smallest geometric unit, which can fully reflect the characteristics of crystals, is usually selected from the lattice to analyze the regularity of atom arrangement in crystals. This smallest geometric unit is called the cell.

Crystal axis

In crystallography, a node at the cell angle is usually taken as the origin and three coordinate axes X, Y and Z are made along its three edges, which are called crystal axes.

The front, back and top of the coordinate origin are defined as the positive direction of the axis, and vice versa. The shape and size of the cell are expressed by the length of edges and the angle between edges. The whole lattice is composed of many cells of the same size, shape and orientation, which are accumulated repeatedly in space.

In our daily life, besides pure metal, what kinds of metal materials are used for us?

We make use of the excellent electrical conductivity, thermal conductivity, chemical stability and beautiful metallic luster of pure metals, which are widely used in various conductors, heat exchangers, decorations, artworks, etc. But almost all kinds of metals have poor mechanical properties such as strength, hardness and wear resistance, so they are not suitable for making mechanical parts and tools with higher requirements for mechanical properties. In addition, the types of pure metals are limited (only about 79) and the price is high, so they are rarely used in industry.

第二章　铁碳合金与钢的热处理

2.1　铁碳合金

合金

两种或两种以上的金属元素，或金属与非金属元素熔合在一起，构成具有金属特性的物质，称为合金。

组元

组成合金的元素称作组元，简称元。如铁、碳是钢和铸铁的组元。合金中的稳定化合物也可作为组元。

相

因为合金组元的相互作用可构成不同的相。在合金中，凡化学成分和晶格构造相同、并与其他部分有界面分开的均匀组成部分称为相。

例如：钢处于液态时，称为液相；但是在结晶过程中，会出现固态和液态共存的现象，此时，它们各为一相。

铁碳合金

铁碳合金是以铁和碳为基本组元的合金。一般将碳的质量分数为 0.02%～2.11% 的铁碳合金称为碳钢，碳的质量分数大于 2.11% 的称为铸铁。

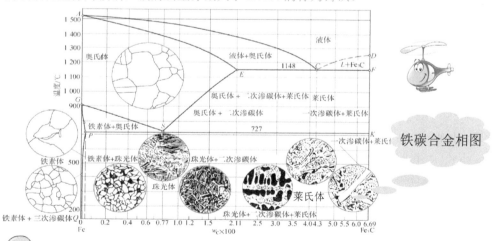

铁碳合金相图

在国家标准规定中，含碳量大于 2% 的铁合金是铸铁。那对钢又是怎么解释的呢？我们为什么要研究钢的应用呢？

钢是对含碳量质量百分比介于 0.02% 至 2.11% 之间的铁碳合金的统称。钢的化学成分可以有很大变化，只含碳元素的钢称为碳素钢（碳钢）或普通钢；在实际生产中，钢往往根据用途的不同含有不同的合金元素，比如锰、镍、钒等。人类直到 19 世纪贝氏炼钢法发明之前，都认为钢的制取是一项高成本低效率的工作。现如今，钢以其低廉的价格、可靠的性能成为世界上使用最多的材料之一，是建筑业、制造业和人们日常生活中不可或缺的成分，可以说钢是现代社会的物质基础。

Alloy

Two or more metal elements, or metal and non-metal elements are fused together to form constitute a material with metal characteristics, known as alloys.

Component

Elements that make up alloys are called components, or elements for short. For example, iron and carbon are components of steel and cast iron. Stable compounds in alloys can also be used as components.

Phase

Because the interaction of alloy components can form different phases. In alloys, homogeneous components with the same chemical composition and lattice structure and separated from other parts are called phases. For example, when steel is in liquid state, it is called liquid phase; but in the process of crystallization, there will be the coexistence of solid and liquid, at this time, they are one phase.

Iron carbon alloy

It is an alloy with iron and carbon as its basic components. Ferrocarbon alloys with carbon content of 0.0218%-2.11% are generally called carbon steel, and iron alloys with carbon content of more than 2.11% are called cast iron.

In the national standard, the ferroalloy with more than 2% carbon content is cast iron. How does that explain steel? Why should we study the application of steel?

Steel is a general term for iron-carbon alloys whose carbon content is between 0.02% and 2.11%. The chemical composition of steel can vary greatly. Steel containing only carbon elements is called carbon steel (carbon steel) or ordinary steel. In actual production, steel often contains different alloying elements according to different uses, such as manganese, nickel, vanadium and so on. Until the invention of Bayesian steelmaking in the 19th century, it is believed that the production of steel was a high-cost and efficient work. Nowadays, steel has become one of the most used materials in the world because of its low price and reliable performance, and it is an indispensable component in the construction industry, manufacturing industry and people's daily life. It can be said that steel is the material basis of modern society.

2.2 钢的热处理

按碳含量高低分类

低碳钢：碳含量一般低于0.25%（质量分数）

中碳钢：碳含量一般为0.25%~0.6%（质量分数）

高碳钢：碳含量一般高于0.6%（质量分数）。

 改善钢的性能有两个主要途径：一是调整钢的化学成分，加入合金元素，即合金化的办法；二是进行钢的热处理，改善钢的组织结构。其中，后一种是现代机械加工生产中经常采用的办法。

钢的热处理

钢的热处理是将钢在固态下，通过加热、保温和冷却，以改变钢的组织，从而获得所需性能的工艺方法。

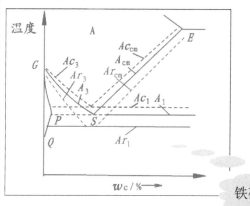

铁碳相图

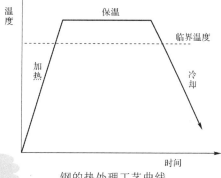

钢的热处理工艺曲线

铁碳相图及钢的临界点是在无限缓慢的加热和冷却条件下测得的。但在实际应用中，加热和冷却都有一定的速度，所以钢中组织的转变总是"落后一拍"，即加热时的实际转变温度总是高于 A_1、A_3 或 A_{cm}，称为加热时的实际临界点，记作 Ac_1、Ac_3 或 Ac_{cm}；冷却时的实际转变温度总是低于 A_1、A_3 或 A_{cm}，称为冷却时的实际临界点，记作 Ar_1、Ar_3 和 Ar_{cm}。

加热时，各实际临界点与平衡临界点的差称为过热度。冷却时，各实际临界点与平衡临界点的差称为过冷度。加热时，过热度与加热速度有关，加热速度愈快，过热度愈大，因而 Ac_1、Ac_3 和 Ac_{cm} 也愈高；相反，冷却时，过冷度与冷却速度有关，冷却速度愈大，过冷度就愈大，则 Ar_1、Ar_3 和 Ar_{cm} 也就愈低。过热度或过冷度愈大，组织的转变就愈滞后。

这么复杂的图，我看不懂呀！

钢的实际临界点的含义如下：

Ac_1——加热时，珠光体转变为奥氏体的温度。

Ac_3——加热时，铁素体全部溶入奥氏体的温度。

Ac_{cm}——加热时，二次渗碳体全部溶入奥氏体的温度。

Ar_1——冷却时，奥氏体转变为珠光体的温度。

Ar_3——冷却时，铁素体从奥氏体中析出的开始温度。

Ar_{cm}——冷却时，二次渗碳体从奥氏体中析出的开始温度。

别着急！我来解释一下！

钢的热处理只改变金属材料的组织和性能，而不改变其形状和尺寸。它可以提高零件的强度、硬度、韧性、弹性，同时，还可以改善毛坯或原材料的切削性能，使之易于加工。

热处理已成为一种不可缺少的工艺方法，在机械制造业中应用广泛。在机床制造中有 60%～70% 的零件需要进行热处理；在汽车、拖拉机中有 70%～80% 的零件需要进行热处理；在各类工具（刃具、模具、量具等）和滚动轴承制造中，几乎 100% 的零件需要热处理。

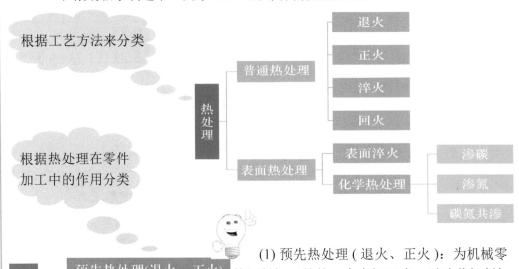

(1) 预先热处理（退火、正火）：为机械零件切削加工前的一个中间工序，以改善切削加工性能及为后续工序作准备。

(2) 最终热处理（淬火、回火）：获得零件最终使用性能的热处理。

		加热	保温	冷却	目的
退火	完全退火	加热到 Ac_3 以上 30℃～50℃	保温	随炉冷到600℃以下，再出炉空气冷却	降低硬度（利于切削或其他种类加工）
	球化退火（主要用于共析钢）	将过共析钢加热到 Ac_1 以上 20℃～30℃		随炉冷到700℃左右，再出炉空气冷却	细化晶粒（提高钢的塑性和韧性）
	去应力退火	将钢加热到 500℃～650℃		随炉冷却	消除内应力（为淬火工序做准备）

	加热	保温	冷却	目的
正火	加热到 Ac_3 或 Ac_{cm} 以上 30℃～50℃		空气中冷却	(1) 去除材料的内应力。 (2) 调整材料的硬度（一般为提高），塑性略降低。这样是为接下来的加工做准备。和退火作用相同，只是为了提高效率，降低成本

	加热	保温	冷却	目的
淬火（水冷淬火、油冷淬火、空冷淬火）	加热到 Ac_3 或 Ac_{c1} 以上 30℃～50℃		快速冷却	提高钢的刚性、硬度、耐磨性、疲劳强度以及韧性等，从而满足各种机械零件和工具的不同使用要求。也可以通过淬火满足某些特种钢材的铁磁性、耐蚀性等特殊的物理、化学性能

		加热	保温	冷却	目的	应用
回火（将淬火后的钢重新加热到 Ac_1 下某温度）	完全退火	工件在 150℃～250℃进行的回火	保温	冷却到室温，在空气或水、油等介质中冷却	保持淬火工件高的硬度和耐磨性，降低淬火残留应力和脆性	各类高碳钢工具、刃具、量具、模具、滚动轴承、渗碳及表面淬火的零件等
	中温回火	工件在 350℃～500℃进行的回火			得到较高的弹性和屈服点，适当的韧性	弹簧、发条、锻模、冲击工具等
	高温回火	工件在 500℃～650℃进行的回火			得到强度、塑性和韧性都较好的综合力学性能	广泛用于各种较重要的受力结构件，如连杆、螺栓、齿轮及轴类零件等

机械加工、成型和安装工艺

◆ 正火与退火的区别：

正火与退火工艺相比，其主要区别是正火的冷却速度稍快，所以正火热处理的生产周期短。故退火与正火同样能达到零件性能要求时，尽可能选用正火。大部分中、低碳钢的坯料一般都采用正火热处理。一般合金钢坯料常采用退火，若用正火，由于冷却速度较快，使其正火后硬度较高，不利于切削加工。

> 退火、正火、淬火、回火是整体热处理中的"四把火"

> 刚刚我们了解了退火与正火工艺性相近，那淬火与回火间又有怎么样的关系呢？

> 马上告诉你

◆ 淬火与回火关系密切，常常配合使用，缺一不可。回火一般紧接着淬火进行，其目的是：

(1) 消除工件淬火时产生的残留应力，防止变形和开裂；
(2) 调整工件的硬度、强度、塑性和韧性，达到使用性能要求；
(3) 稳定组织与尺寸，保证精度；
(4) 改善和提高加工性能。

因此，回火是工件获得所需性能的最后一道重要工序。通过淬火和回火的配合，才可以获得所需的力学性能。

工件淬火并高温回火的复合热处理工艺称为调质。调质不仅作为最终热处理，也可作为一些精密零件或感应淬火件的预先热处理。

表 面 淬 火	
定义	仅使钢铁工件的表面得到淬火的一种表面热处理工艺
目的	提高工件表面的硬度、耐磨性和疲劳强度，而心部仍具有较高的韧性
应用	常用于轴类、齿轮类等零件。表面淬火后一般进行低温回火
分类	根据加热方法不同，可分为感应加热表面淬火、火焰加热表面淬火、电接触加热表面淬火、电解液加热表面淬火等，其中以前两种方法应用最广

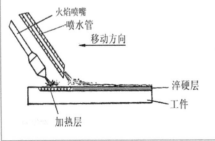

火焰加热表面淬火

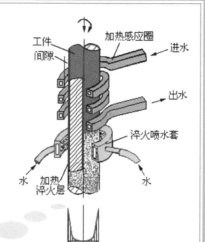

电接触加热表面淬火　　　感应加热表面淬火

	化学热处理
定义	利用化学反应、有时兼用物理方法改变钢件表层化学成分及组织结构，以便得到比均质材料更好的技术、经济效益的金属热处理工艺
目的	使工件表层具有耐磨、耐腐蚀或耐热等性能
应用	由于机械零件的失效和破坏大多数都萌发在表面层，特别在可能引起磨损、疲劳、金属腐蚀、氧化等条件下工作的零件，表面层的性能尤为重要。经化学热处理后的钢件，实质上可以认为是一种特殊复合材料。心部为原始成分的钢，表层则是渗入了合金元素的材料。心部与表层之间是紧密的晶体型结合，它比电镀等表面涂覆技术所获得的心、表部的结合要强得多

钢的渗碳	渗碳就是将低碳钢放入高碳介质中加热、保温，以获得高碳表层的化学热处理工艺
钢的氮化	氮化也称渗氮，是指向工件渗入氮原子，以形成高氮硬化层的化学热处理工艺
钢的碳氮共渗	碳氮共渗是使工件表面同时渗入碳和氮的化学热处理工艺，也称作氰化

　　热处理在机械零件的加工制造过程中有十分重要的地位，与其他工艺过程关系密切。它可以在机械零件加工制造过程中扮演一个重要的中间工序，如改善锻、轧、铸毛坯组织的退火或正火，能消除应力、降低工件硬度、改善切削加工性能；也可以在机械零件性能达到规定技术指标的最终工序中起到重要作用，如经过高温淬火加高温回火，使机械零件获得最为良好的综合机械性能等。

所以说热处理是十分重要的。

Steel Heat Treatment

The heat treatment of steel only changes the structure and properties of metal materials, but does not change their shape and size. It can improve the strength, hardness, toughness and elasticity of parts, and at the same time, it can also improve the cutting performance of blanks or raw materials, making it easy to process. Heat treatment has become an indispensable process, which is widely used in machinery manufacturing industry. In machine tool mechanism, 60% - 70% parts need heat treatment; in automobile and tractor, 70%-80% of the parts need heat treatment; in all kinds of tools (cutting tools, dies, measuring tools, etc.) and rolling bearing manufacturing, almost 100% of the parts need heat treatment.

第二篇 成型加工与压力加工

　　熔炼金属，凝固铸型，塑性变形，这一系列的加工方式使金属材料发生了变化，不单单是物质的种类，而更多应用到了实际的生产生活中。
　　本章介绍金属的铸造、锻造和冲压工艺。

第三章 铸造

3.1 概述

这不是商朝的司母戊方鼎吗？这是铸造出来的吗？

铸造发展的前世今生

我国商朝的司母戊方鼎，战国时期的曾侯乙尊盘，西汉的透光镜，都是古代铸造的代表产品。早期的铸件多为农业生产、宗教、生活等方面的工具或用具，艺术色彩浓厚。我国在公元前513年，铸出了世界上最早见于文字记载的铸铁件——晋国铸型鼎，重约270公斤。欧洲在公元八世纪前后也开始生产铸铁件。在15～17世纪，德、法等国先后敷设了不少向居民供饮用水的铸铁管道。18世纪工业革命以后，蒸汽机、纺织机和铁路等工业兴起，铸件进入为大工业服务的新时期，铸造技术开始有了大的发展。

是的！人类的铸造已有约6000年的历史了，约在公元前1700～前1000年之间我国已进入青铜铸件的全盛期。

进入20世纪，铸造业迅速发展，其重要原因之一是产品技术的进步，即要求铸件的各种机械物理性能要好，还要具有良好的机械加工性能；另一原因是机械工业本身和其他工业例如化工、仪表等的发展，给铸造业创造了有利的物质条件。再者检测手段的发展，保证了铸件质量的提高和稳定，给铸造理论的发展提供了条件，并指导铸造生产。

铸造

铸造——熔炼金属，制造铸型，并将熔融金属浇入铸型，凝固后获得具有一定形状、尺寸和性能的金属零件毛坯的成型方法。

铸造是将金属熔炼成符合一定要求的液体并浇进铸型里，经冷却凝固、清整处理后得到有预定形状、尺寸和性能的铸件的工艺过程。铸造毛坯因接近成型，而达到免于机械加工或少量加工的目的，降低了成本并在一定程度上减少了制作时间。铸造是现代装备制造工业的基础工艺之一。

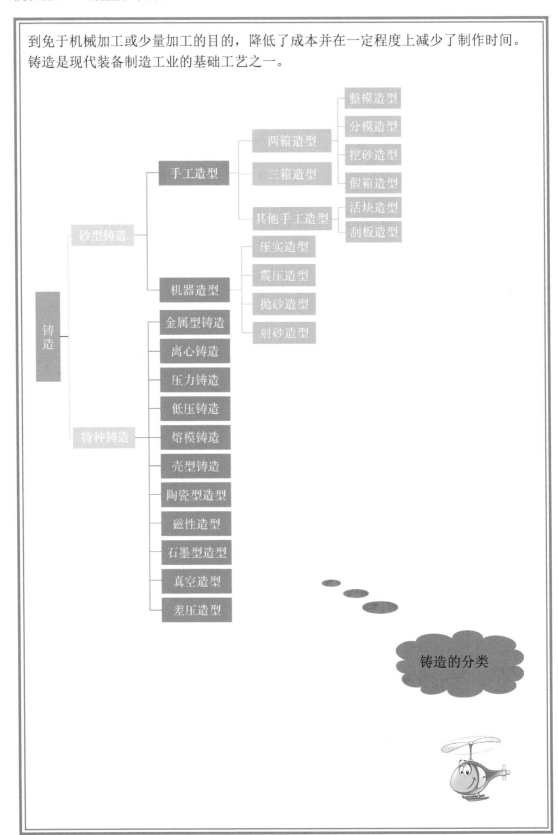

铸造的分类

Metal Casting

Metals are cast into shapes by melting them into a liquid, pouring the metal in a mold, and removing the mold material or casting after the metal has solidified as it cools. The most common metals processed are aluminium and cast iron. However, other metals, such as bronze, brass, steel, magnesium, and zinc, are also used to produce castings in foundries. In this process, parts of desired shapes and sizes can be formed.

Metal Casting assortment
1. Sand casting process
(1) Hand mould
A. Two-part molding
- One-piece patten
- Parted patten
- Oddside molding

B. Three-part molding
C. Others
(2) Machine molding
A. Squeezing molding
B. Impeller ramming
2. Special casting process
(1) Permanent mold casting
(2) Centrifugal casting
(3) Die-casting
(4) Low pressure casting
(5) Fusible patten casting
(6) Shell molding
(7) Ceramic slurry
(8) Magnetic shot molding process
(9) Suction casting
(10) Counter-pressure casting

机械加工、成型和安装工艺

铸造的优点:
(1) 可以生产形状复杂的零件,尤其是复杂内腔的毛坯。
(2) 适应性广,工业常用的金属材料均可铸造,铸件的重量从几克到几百吨都有。
(3) 原材料来源广,价格低廉,如废钢、废件、切屑等。
(4) 铸件的形状尺寸与零件非常接近,减少了切削量,属于无切削加工。
(5) 应用广泛,农业机械中40%～70%、机床中70%～80%的重量都是铸件。

铸造的缺点:
(1) 机械性能不如锻件,组织粗大,缺陷多。
(2) 砂型铸造往往是单件、小批量生产,工人劳动强度大。
(3) 铸件质量不稳定,工序多,影响因素复杂,易产生许多铸造缺陷,对铸件质量有着重要的影响。

以砂型模具为例,加入动画,向学生展示

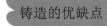
铸造工艺流程图   铸造的优缺点

模具制造
造型(芯)合箱
熔炼浇注
后道处理
完成

第二篇　成型加工与压力加工

铸造模具

预先用其他容易成型的材料做成零件的结构形状即形成模具，然后在砂型中放入模具，于是砂型中就形成了一个和零件结构尺寸一样的空腔，再在该空腔中浇注流动性液体，该液体冷却凝固之后就能形成和模具形状结构完全一样的零件了。铸造模具是铸造工艺中重要的一环。

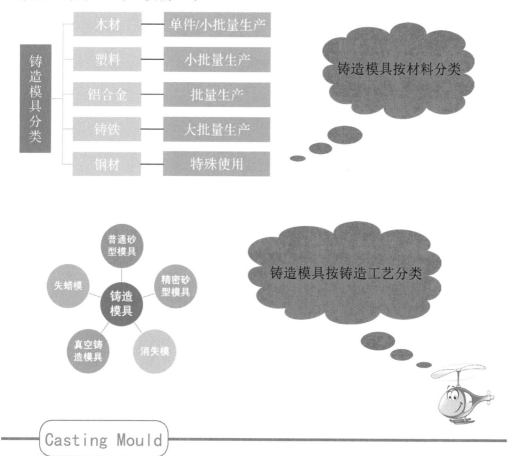

Casting Mould

In order to obtain the structural shape of a part, it is necessary to make the structural shape of the part with other easy-forming materials in advance, and then put the mould into the sand mould, so that a cavity with the same structural size of the part is formed in the sand mould, and then a flowing liquid is poured into the cavity, which can be formed after cooling and solidification. The parts that are exactly the same as the shape and structure of the mold. Casting mold is an important part of casting process.

Classification of Casting Mold:
1. wood: single-piece or small-batch production
2. plastic: small-batch production

3. aluminium alloy: batch production
4. cast iron: mass production
5. steels: special use

Classification of Casting Mold base on the casting technology:
1. sand casting mold
2. lost-wax casting
3. lost foam casting mold
4. Evaporative-pattern casting
5. precision sand mold

3.2 砂型铸造

为什么要介绍砂型铸造呢？

那是由于砂型铸造所用的造型材料价廉易得，铸型制造简便，对铸件的单件生产、成批生产和大量生产均能适应，长期以来，一直是铸造生产中的基本工艺。

砂型铸造

砂型铸造是指在砂型中生产铸件的铸造方法。钢、铁和大多数有色合金铸件都可用砂型铸造的方法获得。传统砂型铸造工艺的基本流程有以下几步：配砂、制模、造芯、造型、浇注、落砂、打磨加工、检验等。

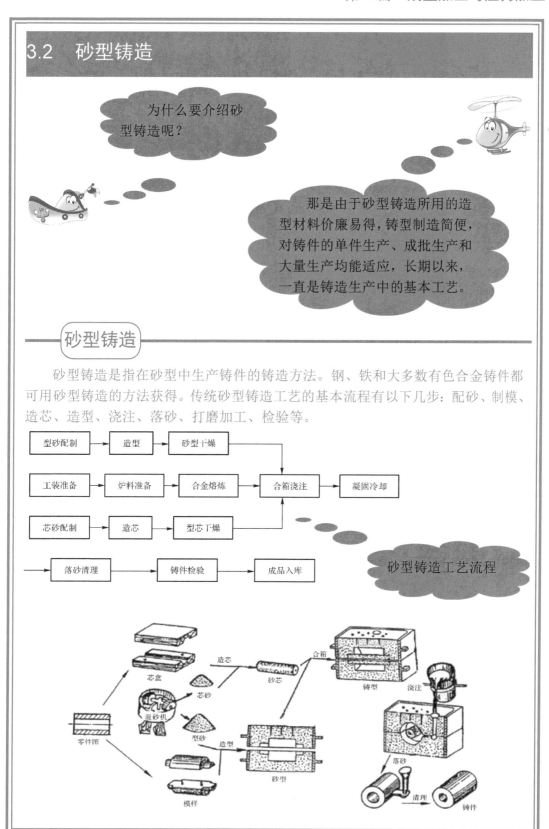

砂型铸造工艺流程

机械加工、成型和安装工艺

(1) 配砂阶段：制备型砂和芯砂，供造型所用。一般使用混砂机放入旧砂和适量黏土进行搅拌。

实际生产过程中是这样的。

配砂阶段

(2) 制模阶段：根据零件图纸制作模具和芯盒，一般单件可以用木模，批量生产可制作塑料模具或金属模（俗称铁模或钢模），大批量铸件可以制作型板。现在模具基本都是用雕刻机，所以制作周期大大缩短，制模一般需要 2～10 天不等。

制模阶段

(3) 造型（制芯）阶段：包括了造型（用型砂形成铸件的型腔）、制芯（形成铸件的内部形状）、配模（把坭芯放入型腔里面，把上下砂箱合好）。造型是铸造中的关键环节。

(4) 熔炼阶段：按照所需要的金属成分配好化学成分，选择合适的熔化炉熔化合金材料，形成合格的液态金属（包括成分合格、温度合格）。熔炼一般采用冲天炉或者电炉（由于环保要求，冲天炉现在基本取缔，通常使用电炉）。

(5) 浇注阶段：用铁水包把电炉里熔化的铁水注入造好的型里。浇注铁水需要注意浇注的速度，使铁水注满整个型腔。

造型阶段

制芯阶段

好危险呀！

熔炼阶段

浇注阶段

(6) 清理阶段：浇注并等熔融金属凝固后，拿锤子去掉浇口并震掉铸件的砂子，然后使用喷砂机进行喷砂，这样铸件表面会显得很干净！对要求不严格的铸件毛坯经过检查基本就可以出厂了。

(7) 铸件加工：对于一些有特别要求的铸件或一些铸造无法达到要求的铸件，可能需要简单加工。一般使用砂轮或磨光机进行加工打磨，去掉毛刺，使铸件更光洁。

浇口、冒口去除

喷砂

(8) 铸件检验：一般在清理或加工阶段过程中，不合格的就已经被挑出来了。但有一些铸件有特殊要求，需要再进行检查。比如有些铸件需要中心孔能插入5厘米的轴，那么就需要拿5厘米的轴进行插入测试。

终于完成啦！

Sand casting, also known as sand molded casting, is a metal casting process characterized by using sand as the mold material. The term "sand casting" can also refer to an object produced via the sand casting process. Sand castings are produced in specialized factories called foundries. Over 60% of all metal castings are produced via sand casting process.

Molds made of sand are relatively cheap, and sufficiently refractory even for steel foundry use. In addition to the sand, a suitable bonding agent (usually clay) is mixed or occurs with the sand. The mixture is moistened, typically with water, but sometimes with other substances, to develop the strength and plasticity of the clay and to make the aggregate suitable for molding. The sand is typically contained in a system of frames or mold boxes known as a flask. The mold cavities and gate system are created by compacting the sand around models called patterns, by carving directly into the sand, or by 3D printing.

There are six steps in this process:

1. Place a pattern in sand to create a mold.
2. Incorporate the pattern and sand in a gating system.
3. Remove the pattern.
4. Fill the mold cavity with molten metal.
5. Allow the metal to cool.
6. Break away the sand mold and remove the casting.

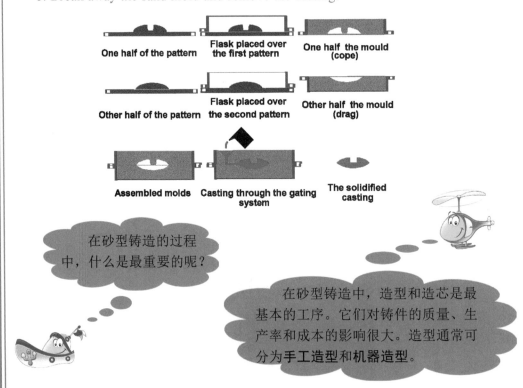

在砂型铸造的过程中，什么是最重要的呢？

在砂型铸造中，造型和造芯是最基本的工序。它们对铸件的质量、生产率和成本的影响很大。造型通常可分为**手工造型和机器造型**。

第二篇 成型加工与压力加工

3.3 造型方法

造型方法	特 点	应用范围
手工造型	用手工或手动工具完成紧砂、起模、修型工序,其特点为: (1) 操作灵活,可按铸件尺寸、形状、批量与现场生产条件灵活地选用具体的造型方法; (2) 工艺适应性强; (3) 生产准备周期短; (4) 生产效率低; (5) 质量稳定性差,铸件尺寸精度、表面质量较差; (6) 对工人技术要求高,劳动强度大	单件、小批量铸件或难以用造型机械生产的形状复杂的大型铸件
机器造型	采用机器完成全部操作,至少完成紧砂操作的造型方法,效率高,铸型和铸件质量好,但投资较大	大量或成批生产的中小铸件

砂型铸造方法及应用范围

手工造型

手工造型是全部用手工或手动工具完成的造型工序。手工造型操作灵活、适应性广、工艺装备简单、成本低,但其铸件质量差、生产率低、劳动强度大、技术水平要求高,所以手工造型主要用于单件小批生产,特别是重型和形状复杂的铸件。

手工造型
- 两箱造型
  - 整模造型
  - 分模造型
  - 挖砂造型
  - 假箱造型
- 三箱造型
- 其他手工造型
  - 活块造型
  - 刮板造型

这么多种造型!难道都是不同款的POSE?嘿嘿~~~~

别开玩笑啦!让我们通过动画来进行了解吧!快来扫一扫啊~~~~

手工造型种类、特点及适用范围

造型方法	特 点	适 用 范 围
整模造型	模样为一整体，分型面为平面，型腔在一个砂箱中，造型方便，不会产生错箱缺陷	铸件最大截面靠一端，且为平直的铸件
分模造型	型腔位于上、下砂箱内，模型制造较复杂，造型方便	最大截面为中部的铸件
挖砂造型	模型是整体的，将阻碍起模的型砂挖掉，分型面是曲面，造型费工	单件小批生产，分型面不是平面的铸件
假箱造型	在造型前预先做出代替底板的底胎，即假箱，再在底胎上做下箱，由于底胎并未参加浇注，故称假箱。假箱造型比挖砂造型操作简单，且分型面整齐	用于成批生产需要挖砂的铸件
三箱造型	中砂箱的高度有一定要求，操作复杂，难以进行机器造型	单件小批生产，中间截面小的铸件
活块造型	将妨碍起模部分做成活块。造型费工，要求操作技术高。活块移位会影响铸件精度	单件小批生产，带有凸起部分又难以起模的铸件
刮板造型	模样制造简化，但造型费工，要求操作技术高	单件小批生产，大、中型回转体铸件

机器造型

 什么是现代化的铸造车间？让铸造生产过程中的造型、制芯、型砂处理、浇注、落砂等工序均由机器来完成，并把这些工艺过程组合成机械化的连续生产流水线，这就形成了现代化的铸造车间。这不仅提高了生产率，而且也提高了铸件精度和表面质量，改善了劳动条件。尽管设备投资较大，但在大批量生产时，铸件成本可显著降低。

将造型过程中的两项最主要的操作——紧砂和起模实现机械化的造型方法称为机器造型。机器造型是采用模板两箱造型。模板是将模样和浇注系统沿分型面与模底板连成一个组合体的专用模具。造型后，模底板形成分型面，模样形成铸型空腔。模底板的厚度不影响铸件的形状和大小。

机器造型按紧实方式的不同分类

压实造型

压实造型是利用压头的压力将砂箱内的型砂紧实。压实造型生产率较高，但存在着在砂箱高度方向上紧实度不够均匀的现象，越接近模底板，紧实度越差，因此，压实造型只适于高度不大的砂箱。

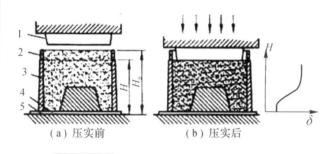

1—压头；
2—辅助框；
3—砂箱；
4—模底板；
5—工作台

震压造型

震压造型是以压缩空气为驱动力，通过震动和撞击对型砂进行紧实。

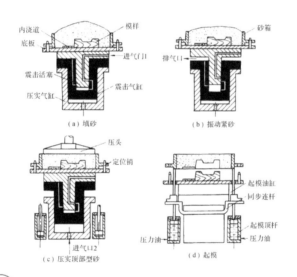

抛砂造型

抛砂头转子上装有叶片，型砂由皮带输送机连续地送入，高速旋转的叶片接住型砂并分成一个个砂团，当砂团随叶片转到出口处时，由于离心力的作用，以高速抛入砂箱，同时完成填砂与紧实。

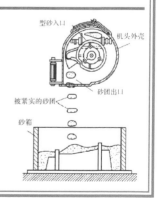

机械加工、成型和安装工艺

抛砂造型

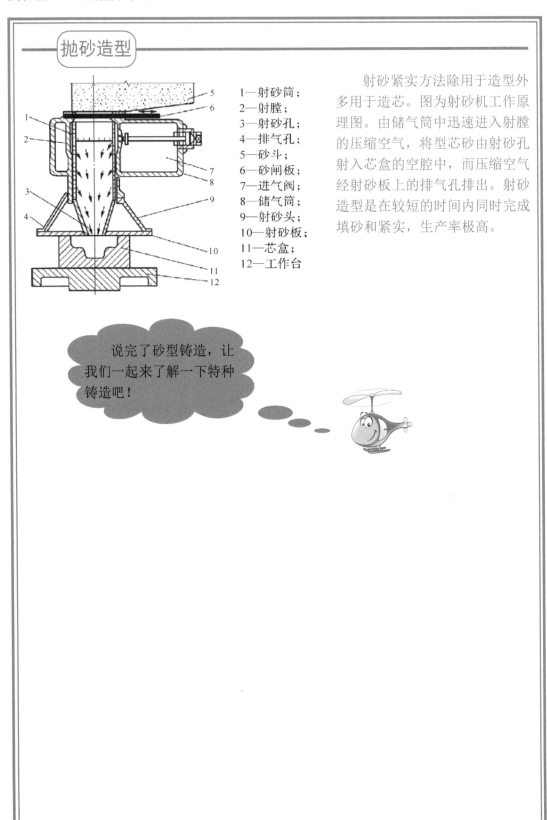

1—射砂筒；
2—射膛；
3—射砂孔；
4—排气孔；
5—砂斗；
6—砂闸板；
7—进气阀；
8—储气筒；
9—射砂头；
10—射砂板；
11—芯盒；
12—工作台

射砂紧实方法除用于造型外多用于造芯。图为射砂机工作原理图。由储气筒中迅速进入射膛的压缩空气，将型芯砂由射砂孔射入芯盒的空腔中，而压缩空气经射砂板上的排气孔排出。射砂造型是在较短的时间内同时完成填砂和紧实，生产率极高。

> 说完了砂型铸造，让我们一起来了解一下特种铸造吧！

3.4 特种铸造

有别于砂型铸造方法的其他铸造工艺,称为特种铸造。特种铸造铸型用砂较少或不用砂,采用特殊工艺装备进行铸造。特种铸造的铸件精度和表面质量高,铸件内在性能好,原材料消耗低,工作环境好。但铸件的结构、形状、尺寸、重量、材料种类往往受到一定限制。特种铸造包括以下几种。

金属型铸造

将液体金属在重力作用下浇入金属铸型以获得铸件的铸造方法,为金属型铸造。铸型用金属制成,可以反复使用几百次到几千次。

离心铸造

离心铸造是指将熔融金属浇入旋转的铸型中,使液体金属在离心力作用下充填铸型并凝固成型的一种铸造方法。

压力铸造

压力铸造是在高压的作用下,以很高的速度把液态或半液态金属压入压铸模型腔,并在压力下快速凝固而获得铸件的铸造方法。

低压铸造

低压铸造是介于一般重力铸造和压力铸造之间的铸造方法。浇注时金属液在低压(20~60 kPa)作用下,由下而上地填充铸型型腔,并在压力下凝固而形成铸件。

熔模铸造

熔模铸造是近净成形、近终成形加工的重要方法之一。熔模铸造又称失蜡铸造、熔模精密铸造、包模精密铸造等,是精密铸造的一种。

壳型铸造

铸造生产中,砂型(芯)直接承受液体金属作用的只是表面一层厚度仅为数毫米的砂壳,其余的砂只起支撑这一层砂壳的作用。若只用一层薄壳来制造铸件,将减少砂处理工步的大量工作,并能减少环境污染。1940 年,Johannes Croning 发明用热法制造壳型,称为"C 法"或"壳法"(shell process),或叫壳型造型 (shell molding)。目前该法不仅可用于造型,更主要的是可用于制壳芯。

机械加工、成型和安装工艺

陶瓷铸造

陶瓷铸造是以耐火度高、热膨胀系数小的耐火材料为骨架材料，用经过水解的硅酸乙酯作为黏结剂而配制成的陶瓷型浆料，在碱性催化剂的作用下，用灌浆法成型，经过胶结、喷燃和烧结等工序，制成光洁、细致、精确的陶瓷型。陶瓷铸造集砂型铸造和熔模铸造的优点，操作及设备简单，型腔尺寸精度高、表面粗糙度低。在单件小批量生产的条件下，可铸造精密铸件，铸件重量从几公斤到几吨；生产率较高，成本低，节省机加工时间。

磁型铸造

磁型铸造是德国在研究消失模铸造的基础上发明的。其实质是采用铁丸代替型砂及型芯砂，用磁场作用力代替铸造黏结剂，用泡沫塑料汽化模代替普通模样的一种新的铸造方法。它的质量状况与消失模铸造相同，但比实型铸造减少了铸造材料的消耗。磁型铸造经常用于自动化生产线上，可铸材料和铸件大小范围较广，常用于汽车零件等精度要求高的中小型铸件生产。

注：消失模铸造（又称实型铸造）是用泡沫塑料(EPS、STMMA或EPMMA)高分子材料制作成与要生产铸造的零件结构、尺寸完全一样的实型模具，经过浸涂耐火材料（起强化、光洁、透气作用）并烘干后，埋在干石英砂中经三维振动造型，浇铸造型砂箱在负压状态下浇入熔化的金属液，使高分子材料模型受热汽化抽出，进而被液体金属取代，冷却凝固后形成一次性成型铸件，是一种新型铸造方法。

石墨型铸造

石墨型铸造是用高纯度的人造石墨块，经机械加工成型或以石墨砂作骨架材料，添加其他附加物制成铸型，浇注凝固后获得铸件的工艺方法。它与砂型、金属型铸造相比，铸型的激冷能力强，能使铸件的晶粒细化，力学性能提高，铸件表面质量好。采用石墨型铸造的铸件受热后的尺寸变化小，且不易发生弯曲、变形，故铸件尺寸精度高。石墨型铸型的使用寿命达2～5万次，其生产率比砂型提高2～10倍。

真空铸造

真空铸造是将结晶器的下端浸入金属液中，抽气使结晶器型腔内造成一定的真空，金属液被吸入型腔一定的高度，受循环水冷却的结晶器产生激冷，金属液由外向内迅速凝固，形成实心或空心的铸件。

反差铸造

反差铸造（差压铸造）的实质是使液态金属在压差的作用下，浇注到预先有一定压力的型腔内，凝固后获得铸件的工艺方法。其特点是充型速度可以控制，铸件充型性好，表面质量高，而且铸件的晶粒细，组织致密，力学性能好。

Permanent Mold Casting

Permanent mold casting is a metal casting process that employs reusable molds ("permanent molds"), usually made from metal. The most common process uses gravity to fill the mold, however gas pressure or a vacuum are also used. A variation on the typical gravity casting process, called slush casting, produces hollow castings. Common casting metals are aluminum, magnesium, and copper alloys. Other materials include tin, zinc, lead alloys and iron and steel are also cast in graphite molds. Typical products are components such as gears, splines, wheels, gear housings, pipe fittings, fuel injection housings, and automotive engine pistons.

Centrifugal Casting

Centrifugal casting or rotocasting is a casting technique that is typically used to cast thin-walled cylinders. It is typically used to cast materials such as metals, glass, and concrete. A high quality is attainable by control of metallurgy and crystal structure. Unlike most other casting techniques, centrifugal casting is chiefly used to manufacture rotationally symmetric stock materials in standard sizes for further machining, rather than shaped parts tailored to a particular end-use.

Die-Casting

Die casting is a metal casting process that is characterized by forcing molten metal under high pressure into a mold cavity. The mold cavity is created using two hardened tool steel dies which have been machined into shape and work similarly to an injection mold during the process. Most die castings are made from non-ferrous metals, specifically zinc, copper, aluminium, magnesium, lead, pewter, and tin-based alloys. Depending on the type of metal being cast, a hot- or cold-chamber machine is used.

The casting equipment and the metal dies represent large capital costs and this tends to limit the process to high-volume production. Manufacture of parts using die casting is relatively simple, involving only four main steps, which keeps the incremental cost per item low. It is especially suited for a large quantity of small- to medium-sized castings, which explains why die casting produces more castings than any other casting process. Die castings are characterized by a very good surface finish (by casting standards) and dimensional consistency.

Low Pressure Casting

Low pressure casting is a casting method between general gravity casting and die casting. Under the action of low pressure (20-60 kPa), molten metal fills the mould cavity from bottom to top and solidifies under pressure to form castings.

Low Pressure Casting

Low pressure casting is a casting method between general gravity casting and die casting. Under the action of low pressure (20-60 kPa), molten metal fills the mould cavity from bottom to top and solidifies under pressure to form castings.

第四章 锻造

4.1 概述

锻造的发展与应用

锻造在中国有着悠久的历史,它是以手工作坊的生产方式延续下来的。大概在20世纪初,锻造才逐渐以机械工业化的生产方式出现在铁路、兵工、造船等行业中。这种转变的主要标志就是使用了锻造能力强大的机器。

在汽车制造过程中,广泛地采用锻造的加工方法。随着科技的进步,对工件精度的要求不断提高,具有高效率、低成本、低能耗、高质量等优点的精密锻造技术得到越来越广泛的应用。近年来,随着精密锻造技术的迅速发展,冷锻件和温锻件越来越多地用到汽车工业中,产品形状越来越接近最终形状。精密锻造将随着未来工艺和相关技术的进步得到相应发展。依据金属塑性成形时的变形温度不同,通过精密冷锻技术生产的汽车零部件包括:汽车离合器接合齿圈、汽车变速器的输入轴零件、轴承圈、汽车等速万向节滑套系列产品、汽车差速器齿轮、汽车前轴等。

锻造

锻造是一种利用锻压机械对金属坯料施加压力,使其产生塑性变形以获得具有一定机械性能、一定形状和尺寸锻件的加工方法。通过锻造能消除金属在冶炼过程中产生的铸态疏松等缺陷,优化微观组织结构,同时由于保存了完整的金属流线,锻件的机械性能一般优于同样材料的铸件。相关机械中负载高、工作条件严峻的重要零件,除形状较简单的可用轧制的板材、型材或焊接件外,多采用锻件。

第二篇　成型加工与压力加工

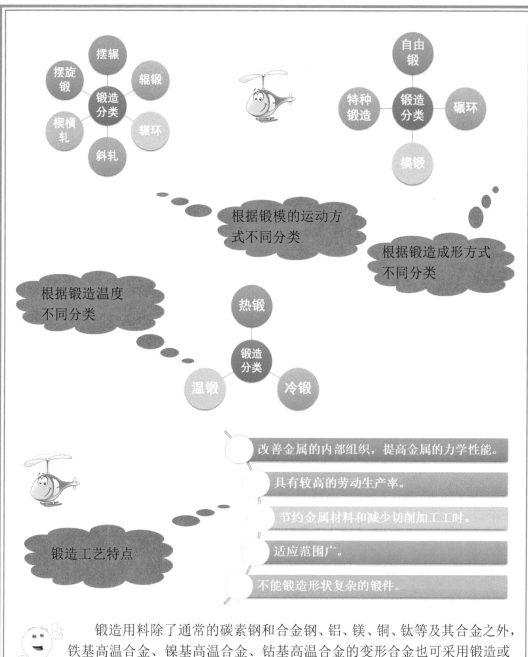

锻造用料除了通常的碳素钢和合金钢、铝、镁、铜、钛等及其合金之外，铁基高温合金、镍基高温合金、钴基高温合金的变形合金也可采用锻造或轧制的方式。但是这些合金由于其塑性区相对较窄，所以锻造难度会相对较大，不同材料的加热温度、开锻温度与终锻温度都有严格的要求。

材料的原始状态有棒料、铸锭、金属粉末和液态金属。金属在变形前的横断面积与变形后的横断面积之比称为锻造比。

正确地选择锻造比、合理的加热温度及保温时间、合理的始锻温度和终锻温度、合理的变形量及变形速度对提高产品质量、降低成本有很大关系。

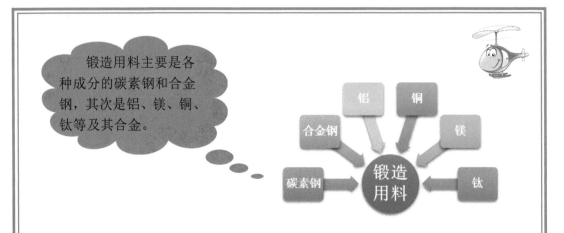

Forging

Forging is a manufacturing process involving the shaping of metal using localized compressive forces. The blows are delivered with a hammer (often a power hammer) or a die. Forging is often classified according to the temperature at which it is performed: cold forging (a type of cold working), warm forging, or hot forging (a type of hot working). For the latter two, the metal is heated, usually in a forge. Forged parts can range in weight from less than a kilogram to hundreds of metric tons.[1][2] Forging has been done by smiths for millennia; the traditional products were kitchenware, hardware, hand tools, edged weapons, cymbals, and jewellery. Since the Industrial Revolution, forged parts are widely used in mechanism s and machines wherever a component requires high strength; such forgings usually require further processing (such as machining) to achieve a finished part. Today, forging is a major worldwide industry.

History

Forging is one of the oldest known metalworking processes. Traditionally, forging was performed by a smith using hammer and anvil, though introducing water power to the production and working of iron in the 12th century allowed the use of large trip hammers or power hammers that exponentially increased the amount and size of iron that could be produced and forged easily. The smithy or forge has evolved over centuries to become a facility with engineered processes, production equipment, tooling, raw materials and products to meet the demands of modern industry.

In modern times, industrial forging is done either with presses or with hammers powered by compressed air, electricity, hydraulics or steam. These hammers may have reciprocating weights in the thousands of pounds. Smaller power hammers, 500 lb (230 kg) or less reciprocating weight, and hydraulic presses are common in art smithies as well. Some steam hammers remain in use, but they became obsolete with the availability of the other more convenient power sources.

Advantages and disadvantages

Forging can produce a piece that is stronger than an equivalent cast or machined part. As the metal is shaped during the forging process, its internal grain texture deforms to follow the general shape of the part. As a result, the texture variation is continuous throughout the part, giving rise to a piece with improved strength characteristics.[4] Additionally, forgings can target a lower total cost when compared to a casting or fabrication. Considering all the costs that are involved in a product's lifecycle from procurement to lead time to rework, and factoring in the costs of scrap, downtime and further quality issues, the long-term benefits of forgings can outweigh the short-term cost-savings that castings or fabrications might offer.

Some metals may be forged cold, but iron and steel are almost always hot forged. Hot forging prevents the work hardening that would result from cold forging, which would increase the difficulty of performing secondary machining operations on the piece. Also, while work hardening may be desirable in some circumstances, other methods of hardening the piece, such as heat treating, are generally more economical and more controllable. Alloys that are amenable to precipitation hardening, such as most aluminium alloys and titanium, can be hot forged, followed by hardening.

Production forging involves significant capital expenditure for machinery, tooling, facilities and personnel. In the case of hot forging, a high-temperature furnace (sometimes referred to as the forge) is required to heat ingots or billets. Owing to the size of the massive forging hammers and presses and the parts they can produce, as well as the dangers inherent in working with hot metal, a special building is frequently required to house the operation. In the case of drop forging operations, provisions must be made to absorb the shock and vibration generated by the hammer. Most forging operations use metal-forming dies, which must be precisely machined and carefully heat-treated to correctly shape the workpiece, as well as to withstand the tremendous forces involved.

Processes

Many different kinds of forging processes are available; however, they can be grouped into three main classes:

1. Drawn out: length increases, cross-section decreases
2. Upset: length decreases, cross-section increases
3. Squeezed in closed compression dies: produces multidirectional flow

Common forging processes include: roll forging, swaging, cogging, open-die forging, impression-die forging, press forging, automatic hot forging and upsetting.

机械加工、成型和安装工艺

4.2 自由锻造

自由锻造是利用冲击力或压力使金属在上下砧面间各个方向自由变形，不受任何限制而获得所需形状及尺寸和一定机械性能的锻件的一种加工方法，简称自由锻。

自由锻的优点：
(1) 所用工具和设备简单，通用性好，成本低。
(2) 与铸造毛坯相比，自由锻消除了缩孔、缩松、气孔等缺陷，使毛坯具有更高的力学性能；
(3) 锻件形状简单，操作灵活；
(4) 在重型机器及重要零件的制造上有特别重要的意义；
(5) 对设备的精度要求低；
(6) 生产周期短。

自由锻的缺点：
(1) 自由锻的生产效率比模型锻造低得多；
(2) 锻件形状简单，尺寸精度较低，表面粗糙；
(3) 工人劳动强度高，并且要求技术水平也高；
(4) 不易实现机械化和自动化。

自由锻造的优缺点

自由锻主要应用于单件、小批量生产，修配以及大型锻件的生产和新产品的试制、模锻件的制坯等方面，是特大型锻件的唯一生产方法。

自由锻造又可分为

手工自由锻造：手工自由锻生产效率低，劳动强度大，仅用于修配或简单、小型、小批锻件的生产。

机器自由锻：在现代工业生产中，机器自由锻已成为锻造生产的主要方法。在重型机械制造中，它具有特别重要的作用，而生产的锻件形状和尺寸主要由操作工的技术水平决定。

第二篇 成型加工与压力加工

自由锻造的主要设备

```
          ┌─ 空气锤
   锻锤 ──┤
          └─ 蒸汽—空气锤
          
          ┌─ 油压机
   液压机─┤
          └─ 水压机
```

空气锤

油压机

　　自由锻造的设备分为锻锤和液压机两大类。生产中使用的锻锤有空气锤和蒸汽-空气锤，有些厂还使用结构简单、投资少的弹簧锤、夹板锤、杠杆锤和钢丝锤等。液压机是以液体产生的静压力使坯料变形的，是生产大型锻件的唯一方式。

　◆**锻锤**是以冲击力使坯料变形的，设备规格以落下部分的重量来表示。

　◆**空气锤**的吨位较小，只有 $0.5 \sim 10$ kN，用于锻 100 kg 以下的锻件

　◆**蒸汽-空气锤**的吨位较大，可达 $10 \sim 50$ kN，可锻 1500kg 以下的锻件。

　◆**液压机**是以液体产生的静压力使坯料变形的，设备规格以最大压力来表示。

　◆**水压机**的压力大，可达 $5000 \sim 15000$ kN，是锻造大型锻件的主要设备。

介绍了这么多，那自由锻造要求工人做哪些工作呢？

这其实是自由锻造的工序，下面给大家介绍一下吧！

机械加工、成型和安装工艺

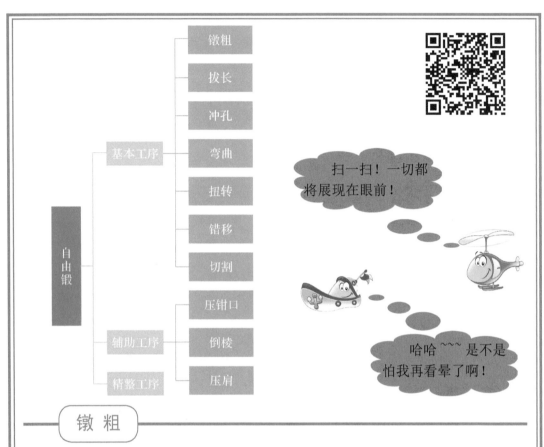

镦 粗

镦粗是使坯料高度减小而横截面增大的锻造工序。若使坯料局部截面增大则称为局部镦粗。镦粗的作用包括由横截面较小的坯料得到横截面较大而高度较小的锻件。镦粗工序主要用于锻造齿轮坯、圆饼类锻件,其可有效地改善坯料组织,减小力学性能的异向性。

镦粗的方法和途径

序号	名　称	简　图	用　途
1	平砧间镦粗		用于镦粗棒料和切去冒口和底部的锭料
2	在带孔的垫球间镦粗		用于锻造带凸座的齿轮、突缘等锻件。当锻件直径较大、凸座直径很小,而且所用的毛坯直径比凸座的直径要大得多时采用
3	在漏盘或模子内局部镦粗		用于锻造带凸座的齿轮和长杆类锻件的头部和凸缘等。这时凸座的直径和高度都较大

镦粗时应注意镦粗部分的长度与直径之比应小于 2.5，否则容易镦弯；坯料端面要平整且与轴线垂直，锻打用力要正，否则容易锻歪；镦粗力要足够大，否则会形成细腰形或夹层。镦粗时要考虑金属塑性的高低，控制其变形程度，对于硬度高的金属，强烈镦粗会产生纵向裂纹。

拔 长

使坯料横截面减小而长度增加的锻造工序称为拔长。拔长主要用于轴杆类锻件成形，其作用是改善锻件内部质量。拔长的种类有平砧铁拔长、芯轴拔长、芯轴扩孔等。

冲 孔

采用冲子将坯料冲出透孔或不透孔的锻造工序叫冲孔。

冲孔方法

序号	冲孔方法	简图	应用范围和工艺参数
1	实心冲子冲孔（双面冲孔）		用于冲一般的孔的工艺参数 (1) $\dfrac{D_0}{d_1} \geq 2.5 \sim 3$ (2) $H_0 \leq D_0$ D_0——原毛坯直径； H_0——原毛坯高度； d_0——冲头直径
2	在垫环上冲孔（漏孔）		用于冲较薄的毛坯 例如锻件高度 H_0 和直径的比值 $\dfrac{H_0}{D} < 0.125$ 时，常采用此法

   据冲孔所用的冲子的形状不同，冲孔分实心冲子冲孔和空心冲子冲孔。实心冲子冲孔分单面冲孔和双面冲孔。

单面冲孔　　　　　双面冲孔

弯 曲

用外形一定的工模具将坯料弯成所规定的外形的锻造工序，称为弯曲。
常用的弯曲方法有以下两种：
(1) 锻锤压紧弯曲法。坯料的一端被上、下砧压紧，用大锤打击或用吊车拉另一端，使其弯曲成形。

(2) 弯曲法。在垫模中弯曲能得到形状和尺寸较准确的小型锻件。

切割

切割是指将坯料分成几部分或部分地割开，或从坯料的外部割掉一部分，或从内部割出一部分的锻造工序。

错移

错移是指将坯料的一部分相对另一部分平行错开一段距离，但仍保持轴心平行的锻造工序，常用于锻造曲轴零件。

错移时，先对坯料进局部切割，然后在切口两侧分别施加大小相等、方法相反且垂直于轴线的冲击力或压力，使坯料实现错移。

扭转

扭转是指将坯料的一部分相对于另一部分绕其轴线旋转一定角度的锻造工序。该工序多用于锻造多拐曲轴和校正某些锻件。小型坯料扭转角度不大时，可用锤击方法。

辅助工序

辅助工序是指在坯料进入基本工序前预先变形的工序。如钢锭倒棱、预压钳把、阶梯轴分段压痕等工步。

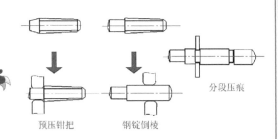

辅助工序

预压钳把　　钢锭倒棱　　分段压痕

修整工序

修整工序是指用来精整锻件尺寸和形状，使其完全达到锻件图纸要求的工序，一般是在某一基本工步完成后进行。如镦粗后的鼓形滚圆和截面滚圆、端面平整、弯曲校直等工步。

介绍完了自由锻造，下面来介绍一下胎模锻吧！

修整工序

鼓形滚圆　　端面平整

弯曲校正

Drop forging

Drop forging is a forging process where a hammer is raised and then "dropped" onto the workpiece to deform it according to the shape of the die. There are two types of drop forging: open-die drop forging and closed-die drop forging. As the names imply, the difference is in the shape of the die, with the former not fully enclosing the workpiece, while the latter does.

Open-die drop forging

Open-die forging is also known as smith forging. In open-die forging, a hammer strikes and deforms the workpiece, which is placed on a stationary anvil. Open-die forging gets its name from the fact that the dies (the surfaces that are in contact with the workpiece) do not enclose the workpiece, allowing it to flow except where contacted by the dies. The operator therefore needs to orient and position the workpiece to get the desired shape. The dies are usually flat in shape, but some have a specially shaped surface for specialized operations. For example, a die may have a round, concave, or convex surface or be a tool to form holes or be a cut-off tool. Open-die forgings can be worked into shapes which include discs, hubs, blocks, shafts (including step shafts or with flanges), sleeves, cylinders, flats, hexes, rounds, plate, and some custom shapes. Open-die forging lends itself to short runs and is appropriate for art smithing and custom work. In some cases, open-die forging may be employed to rough-shape ingots to prepare them for subsequent operations. Open-die forging may also orient the grain to increase strength in the required direction.

Advantages of open-die forging

1. Reduced chance of voids
2. Better fatigue resistance
3. Improved microstructure
4. Continuous grain flow
5. Finer grain size
6. Greater strength

"Cogging" is the successive deformation of a bar along its length using an open-die drop forge. It is commonly used to work a piece of raw material to the proper thickness. Once the proper thickness is achieved the proper width is achieved via "edging". "Edging" is the process of concentrating material using a concave shaped open-die. The process is called "edging" because it is usually carried out on the ends of the workpiece. "Fullering" is a similar process that thins out sections of the forging using a convex shaped die. These processes prepare the workpieces for further forging processes.

Equipment

The most common type of forging equipment is the hammer and anvil. Principles behind the hammer and anvil are still used today in *drop-hammer* equipment. The principle behind the machine is simple: raise the hammer and drop it or propel it into the workpiece, which rests on the anvil. The main variations between drop-hammers are in the way the hammer is powered; the most common being air and steam hammers. Drop-hammers usually operate in a vertical position. The main reason for this is excess energy (energy that isn't used to deform the workpiece) that isn't released as heat or sound needs to be transmitted to the foundation. Moreover, a large machine base is needed to absorb the impacts.

To overcome some shortcomings of the drop-hammer, the *counterblow machine or impactor* is used. In a counterblow machine, both the hammer and anvil move and the workpiece is held between them. Here excess energy becomes recoil, which allows the machine to work horizontally and have a smaller base. Other advantages include less noise, heat and vibration. It also produces a distinctly different flow pattern. Both of these machines can be used for open-die or closed-die forging.

The process of Open-Die Drop Forging:
The Basic Process:
Upsetting
Stretching
Punching
Bending
Twisting
Offset
Cutting
The Auxiliary Process:
Chamfering
Necking
The equipment is include:
Forging hammer
Hydraulic press (air dir forging hammer, steam or air dir forging hammer)

4.3 胎模锻

胎模锻

胎模锻是采用自由锻的方法制坯,然后在胎模中最后成形的一种锻造方法,是介于自由锻与模锻之间的一种锻造方法。胎模锻通常是在自由锻锤或压力机上安装一定形状的模具进行模锻件加工的。它是为了适应中小批量锻件生产而发展起来的一种锻造工艺,兼具有模锻和自由锻的特点。

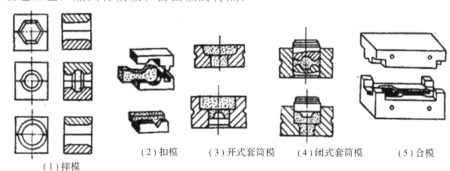

(1)摔模　(2)扣模　(3)开式套筒模　(4)闭式套筒模　(5)合模

胎模锻工序

胎模锻与自由锻造相比有哪些优缺点呢?

胎模锻的优点:
(1) 由于坯料在模膛内成形,所以锻件尺寸比较精确,表面比较光洁,流线组织的分布比较合理,质量较高;
(2) 胎模锻能锻出形状比较复杂的锻件;
(3) 由于锻件形状由模膛控制,所以坯料成形较快,生产率比自由锻高 1~5 倍;
(4) 余块少,因而加工余量较小,既可节省金属材料,又能减少机加工工时。

胎模锻的缺点:
(1) 需要吨位较大的锻锤;
(2) 只能生产小型锻件;
(3) 胎模的使用寿命较低;
(4) 工作时一般要靠人力搬动胎模,因而劳动强度较大;
(5) 胎模锻用于生产中、小批量的锻件。

机械加工、成型和安装工艺

名称	结构和用途
摔模	摔模由上摔、下摔及摔把组成，常用于回转体轴类锻件的成形或精整，或为合模制坯
弯模	弯模由上模、下模组成，用于吊钩、吊环等弯杆类锻件的成形或为合模制坯
合模	合模由上模、下模及导向装置组成，多用于连杆、拨叉等形状较复杂的非回转体锻件终锻成形
扣模	扣模由上扣、下扣组成，有时仅有下扣，主要用于非回转体锻件的整体、局部成形或为合模制坯
冲切模	冲切模由冲头和凹模组成，用于锻件锻后冲孔和切边
组合套模	组合套模由模套及上模、下模组成，用于齿轮、法兰盘等盘类零件的成形

胎模锻锻模的种类、结构及用途？

猜你会有此要求，早已为你准备好啦！

快点让我扫一扫吧！我又不明白啦！

Impression-die forging

Impression-die forging is also called "closed-die forging". In impression-die forging, the metal is placed in a die resembling a mold, which is attached to an anvil. Usually, the hammer die is shaped as well. The hammer is then dropped on the workpiece, causing the metal to flow and fill the die cavities. The hammer is generally in contact with the workpiece on the scale of milliseconds. Depending on the size and complexity of the part, the hammer may be dropped multiple times in quick succession. Excess metal is squeezed out of the die cavities, forming what is referred to as "flash". The flash cools more rapidly than the rest of the material; this cool metal is stronger than the metal in the die, so it helps prevent more flash from forming. This also forces the metal to completely fill the die cavity. The flash is removed after forging.

In commercial impression-die forging, the workpiece is usually moved through a series of cavities in a die to get from an ingot to the final form. The first impression is used to distribute the metal into the rough shape in accordance to the needs of later cavities; this impression is called an "edging", "fullering", or "bending" impression. The following cavities are called "blocking" cavities, in which the piece is working into a shape that more closely resembles the final product. These stages usually impart the workpiece with generous bends and large fillets. The final shape is forged in a "final" or "finisher" impression cavity. If there is only a short run of parts to be done, then it may be more economical for the die to lack a final impression cavity and instead machine the final features. Impression-die forging has been improved in recent years through increased automation which includes induction heating, mechanical feeding, positioning and manipulation, and the direct heat treatment of parts after forging. One variation of impression-die forging is called "flashless forging", or "true closed-die forging". In this type of forging, the die cavities are completely closed, which keeps the workpiece from forming flash. The major advantage of process is that less metal is lost to flash. Flash can account for 20 % to 45% of the starting material. The disadvantages of this process include additional cost due to a more complex die design and the need for better lubrication and workpiece placement. There are other variations of part formation that integrate impression-die forging. One method incorporates casting a forging preform from liquid metal. The casting is removed after it has solidified, but still hot. It is then finished in a single cavity die. The flash is trimmed, then the part is quench hardened. Another variation follows the same process as outlined above, except the preform is produced by the spraying deposition of metal droplets into shaped collectors (similar to the Osprey process).

Closed-die forging has a high initial cost due to the creation of dies and required design work to make working die cavities. However, it has low recurring costs for each part, thus forgings become more economical with greater production volume. This is one of the major reasons closed-die forgings are often used in the automotive and tool industries. Another reason forgings are common in these industrial sectors is that forgings generally have about a 20 percent higher strength-to-weight ratio compared to cast or machined parts of the same material.

4.4 模型锻造

像介绍胎模锻一样，我们先了解一下模型锻造与自由锻造、胎模锻相比的优缺点吧！

好的！听你的！

模型锻造的优点：
(1) 模锻可以锻制形状较为复杂的锻件。
(2) 锻件的形状和尺寸较准确，表面质量好，材料利用率和生产效率高。

模型锻造的缺点：
(1) 模锻需采用专用的模锻设备和锻模，投资大、前期准备时间长。
(2) 由于模锻受三向压应力变形，变形抗力大。

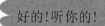

模型锻造简称模锻，是指在专用模锻设备上利用模具使毛坯成形而获得锻件的锻造方法。此方法生产的锻件尺寸精确，加工余量较小，结构也比较复杂，生产率高，适用于中小型锻件的大批量生产。

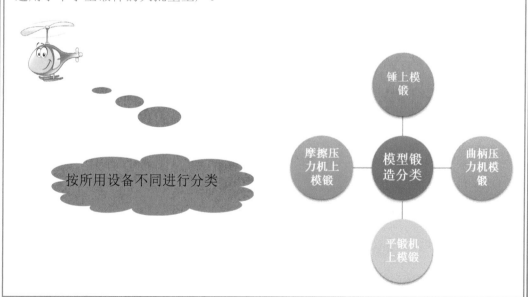

锤上模锻

锤上模锻是将上模固定在锤头上，下模紧固在模垫上，通过随锤头作上下往复运动的上模，对置于下模中的金属坯料施以直接锻击，来获取锻件的锻造方法。锤上模锻是在自由锻和胎模锻的基础上发展起来的一种锻造方法。

下面我们着重介绍一下锤上模锻！

锻模的结构：
- 模锻模膛
  - 预锻模膛
  - 终锻模膛
- 制坯模膛
  - 拔长
  - 滚压
  - 弯曲
  - 切断

锻模的结构

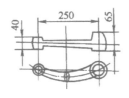

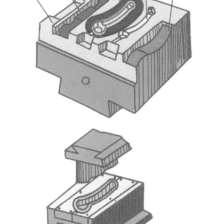

拔长模槽、滚挤模槽、终锻模槽、预锻模槽、弯曲模槽、切断模

原始料坯、延伸、滚挤、弯曲、预锻、终锻、飞边、锻件

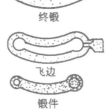

这还是挺有意思的呀！不过，都是什么呢？

第二篇　成型加工与压力加工

预锻模膛

- 目的是使坯料变形到接近于锻件的形状和尺寸，以便在终锻成形时金属充型更加容易，同时减少终锻模膛的磨损，延长锻模的使用寿命。
- 预锻模膛的圆角、模锻斜度均比终锻模膛大，而且不设飞边槽。

终锻模膛

- 可使坯料变形到热锻件所要求的形状和尺寸。待冷却收缩后即达到冷锻件的形状和尺寸。
- 终锻模膛的分模面上有一圈飞边槽，用以增加金属从模膛中流出的阻力，促使金属充满模膛，同时容纳多余的金属。模锻件的飞边须在终锻后切除。

拔长模膛

- 它是用来减少毛坯某部分的横截面积，以增加该部分的长度。拔长模膛分为开式和闭式两种。

滚压模膛

- 它是用来减少毛坯某一部分的横截面积，以增加另一部分的横截面积，从而使金属按锻件形状来分布。滚压模膛分为开式和闭式两种。

弯曲模膛

- 对于弯曲的杆类模锻件，需用弯曲模膛来弯曲毛坯。

切断模膛

- 它是在上模与下模的角上组成一对刀口，用来切断金属。

不怕你不明白，就怕你不扫一扫！

说了这么多，那模型锻造的优缺点又是什么呢？

模锻的优点：

(1) 生产效率较高。模锻时，金属的变形在模膛内进行，故能较快获得所需形状；

(2) 能锻造形状复杂的锻件，并可使金属流线分布更为合理，提高零件的使用寿命；

(3) 模锻件的尺寸较精确，表面质量较好，加工余量较小；

(4) 节省金属材料，减少切削加工工作量；

(5) 在批量足够的条件下，能降低零件成本。

模锻的缺点：

(1) 模锻件的重量受到一般模锻设备能力的限制，大多在 70 kg 以下；

(2) 锻模的制造周期长，成本高；

(3) 模锻设备的投资费用比自由锻大。

锤上模锻设备

蒸汽-空气模锻锤

锤上模锻常用设备

无砧座锤

高速锤

使用最为广泛的设备为：蒸汽-空气锤

第五章 冲 压

5.1 概述

冲压技术是一种具有悠久历史的加工方法和生产制造技术。根据文献记载和考古文物证明，我国古代的冲压加工技术走在世界前列，对人类早期文明社会的进步发挥了重要的作用，作出了重要贡献。利用冲压机械和冲压模具进行的现代冲压加工技术已经有二百年的发展历史。1839年英国成立了Schubler公司，这是早期颇具规模的、现今也是世界上最先进的冲压公司之一。从学科角度上看，到本世纪10年代，冲压加工技术已经从一种从属于机械加工或压力加工工艺的地位，发展成为了一门具有自己理论基础的应用技术科学。俄罗斯（从前苏联时期开始）就有各类冲压技术学校，日本也有冲压工学之说，中国也有冲压工艺学、薄板成形理论方面的教材及专著。可以认为这一学科现已形成了比较完整的知识结构系统。到目前为止，各种不同的关于"未来冲压厂"的话题正逐步趋于一致，这主要是出于资金费用的考虑，希望有比目前性能更好的高性能的冲压生产设备。这种冲压生产线，今天已经不仅用于汽车生产企业，而且在一些汽车部件供应商的企业中也可见到。其中，更换冲压设备生产的产品、增加产品品种已经成为衡量冲压设备的设计和集成能力是否出众的重要因素。今天，对现有设备按照产品进行技术改造并快速投入生产的要求比以往任何时候都更加紧迫。在这些技术工作中，一定范围内的"专用特殊部件"是非常有好处的。在这些所谓的专用特殊部件中，冲压模具仍然是冲压设备最重要的部件并且仍将决定着冲压生产过程。

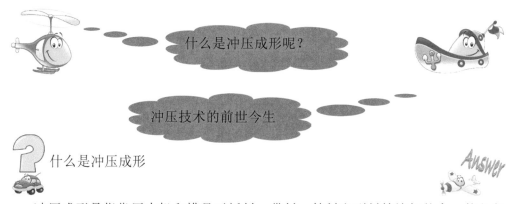

冲压成形是指靠压力机和模具对板材、带材、管材和型材等施加外力，使之产生塑性变形或分离，从而获得所需形状和尺寸的工件（冲压件）的加工成形方法。冲压的坯料主要是热轧和冷轧的钢板和钢带。全世界的钢材中，有60%～70%是板材，其中大部分经过冲压制成成品。汽车的车身、底盘、油箱、散热器片，锅炉的汽包，容器的壳体，电机、电器的铁芯硅钢片等都是冲压加工的。仪器仪表、家用电器、自行车、办公机械、生活器皿等产品中，也有大量冲压件。

机械加工、成型和安装工艺

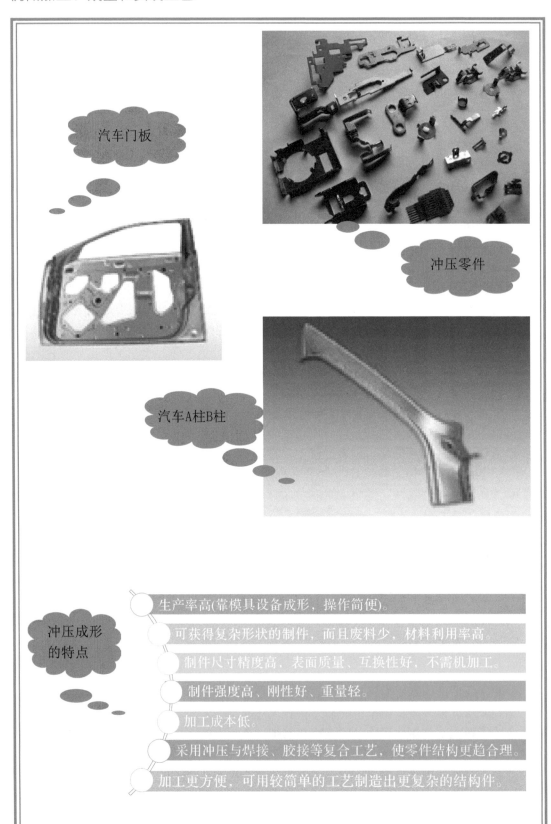

汽车门板

冲压零件

汽车A柱B柱

冲压成形的特点
- 生产率高(靠模具设备成形,操作简便)。
- 可获得复杂形状的制件,而且废料少,材料利用率高。
- 制件尺寸精度高、表面质量、互换性好,不需机加工。
- 制件强度高、刚性好、重量轻。
- 加工成本低。
- 采用冲压与焊接、胶接等复合工艺,使零件结构更趋合理。
- 加工更方便,可用较简单的工艺制造出更复杂的结构件。

Punching

Punching is a forming process that uses a punch press to force a tool, called a punch, through the workpiece to create a hole via shearing. Punching is applicable to a wide variety of materials that come in sheet form, including sheet metal, paper, vulcanized fibre and some forms of plastic sheet. The punch often passes through the work into a die. A scrap slug from the hole is deposited into the die in the process. Depending on the material being punched this slug may be recycled and reused or discarded.

Punching is often the cheapest method for creating holes in sheet materials in medium to high production volumes. When a punch with a special shape is used to create multiple usable parts from a sheet of material the process is known as blanking. In metal forging applications the work is often punched while hot, and this is called hot punching. Slugging is the operation of punching in which punch is stopped as soon as the metal fracture is complete and metal is not removed but held in hole

Punching characteristics

The characteristics of punching are:
1. It is the most cost effective process of making holes in strip or sheet metal for average to high fabrication.
2. It is able to create multiple shaped holes.
3. Punches and dies are usually fabricated from conventional tool steel or carbides?
4. It creates a burnished region roll-over, and die break on sidewall of the resulting hole.
5. It is a quick process.

机械加工、成型和安装工艺

5.2 冲压设备

按传动结构分
- 手动冲床
- 机械冲床
- 液压冲床
- 气动冲床
- 高速机械冲床
- 数控冲床

按加工精度分
- 普通冲床
- 精密冲床

按使用范围分
- 普通冲床
- 专用冲床

冲压设备分类

油压机

手动冲床

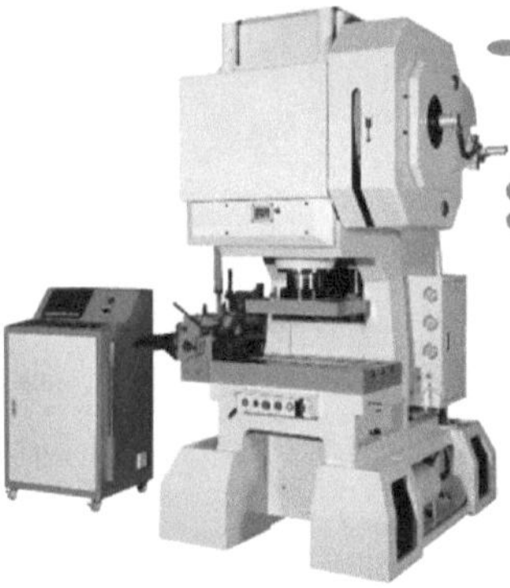

高速冲床

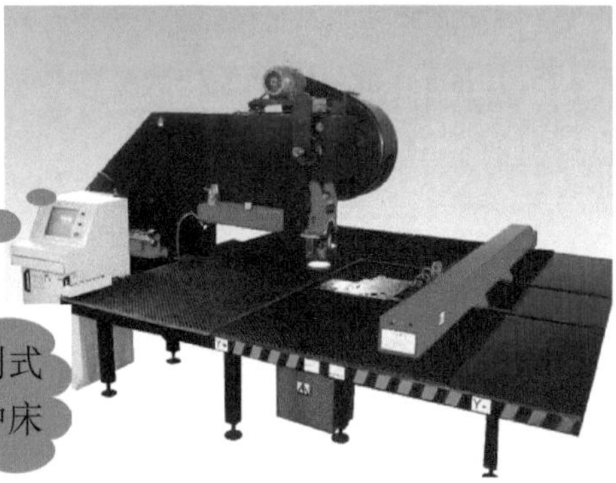

全自动直列式数控转塔冲床

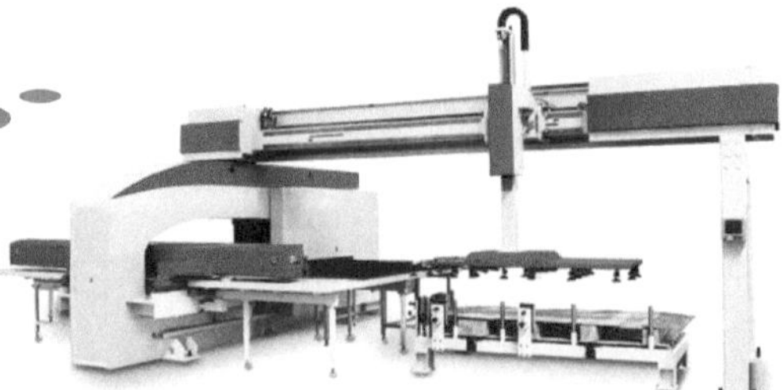

数控板材冲压柔性加工上料系统

Equipment

The classification of equipment by transmission construction:
Manual punching machine
Mechanical punching machine
Hydraulic punching machine
Pneumatic punching machine
High speed mechanical punching machine
CNC punching machine

The classification of equipment by machining precision :
conventional press
Precision press

The classification of equipment by using:
Conventional press
Special press

5.3 冲压的基本工序

冲压主要是按工艺分类，可分为分离工序和成形工序两大类。

◆ 分离工序也称冲裁，其目的是使冲压件沿一定轮廓线从板料上分离，同时保证分离断面的质量要求。

◆ 成形工序的目的是使板料在不破坏的条件下发生塑性变形，制成所需形状和尺寸的工件。在实际生产中，常常是多种工序综合应用于一个工件。冲裁、弯曲、剪切、拉深、胀形、旋压、矫正是几种主要的冲压工艺。

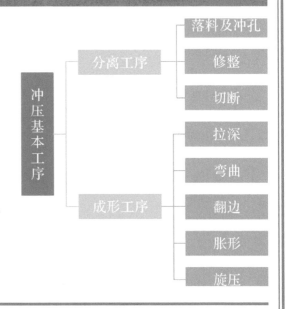

分离工序

使坯料的一部分与另一部分沿一定的轮廓线相互分离的工序，称作分离工序。

◆ 落料及冲孔：落料及冲孔统称为冲裁，冲裁是使坯料按封闭轮廓分离的工序。落料时，冲落部分为成品，而余料是废料。冲孔时，冲落部分是废料，余料部分是成品。

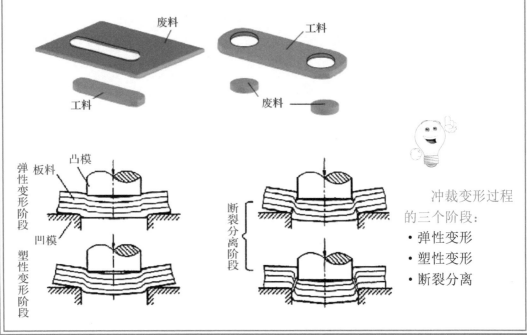

冲裁变形过程的三个阶段：
- 弹性变形
- 塑性变形
- 断裂分离

◆切断：使板料沿不封闭轮廓分离的工序。

◆修整：利用修整模沿冲裁件的外缘或内孔刮去一层薄薄的切屑，以提高冲裁件的加工精度和断面光洁度的冲压方法。

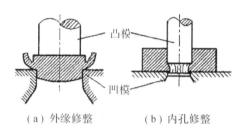

（a）外缘修整　　（b）内孔修整

成形工序

使坯料的一部分相对于另一部分产生位移而不破裂的工序。

◆拉深：指利用拉深模使冲裁后得到的平板坯料变形成开口空心件的工序。可制成筒形、阶梯形、盒形、球形、锥形及其他复杂形状的薄壁零件。

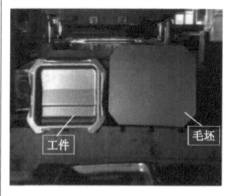

拉深的特点：

(1) 凸模和凹模的特点：与冲裁模不同，它们都有一定的圆角而不是锋利的刃口，其间隙一般稍大于板料厚度。

(2) 变形特点：拉深件的底部一般不变形，厚度基本不变，直壁厚度有所减小。

拉深过程中最容易出现的质量问题：

(1) 拉裂：原因是毛坯被强制拉入凹模时，拉应力超过了材料本身的强度极限。
(2) 起皱：原因是毛坯边缘受压应力过大。

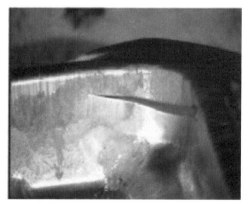

拉裂

起皱

◆ 弯曲：坯料的一部分相对于另一部分弯曲成一定角度的工序。坯料内侧受压，外侧受拉。弯曲时，尽可能使弯曲线与坯料纤维方向垂直以免破裂。

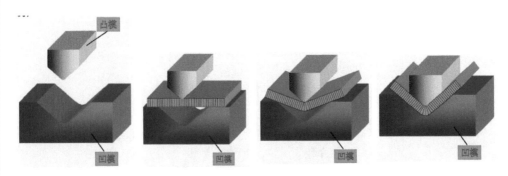

弯曲的特点：

从上图的网格线前后的变化可以看出，弯曲变形有以下特点：
(1) 变形仅仅发生在与凸模接触的圆角范围内。
(2) 坯料内侧受压缩，外侧受拉伸。

◆ 翻边：在坯料的平面部分或曲面部分上，使板料沿一定的曲率翻成竖立边缘的冲压成形工序。

分类：内孔翻边、不变薄翻边、外缘翻边、变薄翻边。

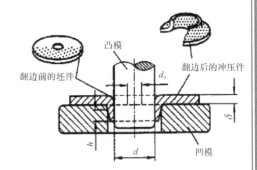

◆ 胀形：主要用于平板毛坯的局部胀形（又叫起伏成形）。
压制凹坑、加强筋、花纹、标记等。
胀形时，毛坯受两向拉应力作用，在此状态下不会产生失稳起皱现象，使得零件表面光滑、质量好。

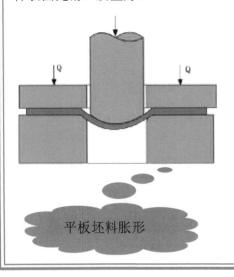

平板坯料胀形

管坯胀形

◆ 旋压。旋压的基本要点：

(1) 合理的转速。
(2) 合理的过渡形状。
(3) 合理加力。

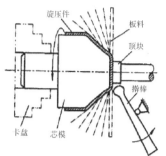

要想完成这些工序，是不是还需要模具呀？

是的，马上为你介绍！在学新知识之前，别忘记扫一扫哦！

Stamping

(also known as pressing) is the process of placing flat sheet metal in either blank or coil form into a stamping press where a tool and die surface forms the metal into a net shape. Stamping includes a variety of sheet-metal forming manufacturing processes, such as punching using a machine press or stamping press, blanking, embossing, bending, flanging, and coining.[1]This could be a single stage operation where every stroke of the press produces the desired form on the sheet metal part, or could occur through a series of stages. The process is usually carried out on sheet metal, but can also be used on other materials, such as polystyrene. Progressive dies are commonly feed from a coil of steel, coil reel for unwinding of coil to a straightener to level the coil and then into a feeder which advances the material into the press and die at a predetermined feed length. Depending on part complexity, the number of stations in the die can be determined.

Stamping is usually done on cold metal sheet. See Forging for hot metal forming operations.

Operations

1. Bending - the material is deformed or bent along a straight line.

2. Flanging - the material is bent along a curved line.

3. Embossing - the material is stretched into a shallow depression. Used primarily for adding decorative patterns. See also Repoussé and chasing.

4. Blanking - a piece is cut out of a sheet of the material, usually making a blank for further processing.

5. Coining - a pattern is compressed or squeezed into the material. Traditionally used to make coins.

6. Drawing - the surface area of a blank is stretched into an alternate shape via controlled material flow. See also deep drawing.

7. Stretching - the surface area of a blank is increased by tension, with no inward movement of the blank edge. Often used to make smooth auto body parts.

8. Ironing - the material is squeezed and reduced in thickness along a vertical wall. Used for beverage cans and ammunition cartridge cases.

9. Reducing/Necking - used to gradually reduce the diameter of the open end of a vessel or tube.

10. Curling - deforming material into a tubular profile. Door hinges are a common example.

11. Hemming - folding an edge over onto itself to add thickness. The edges of automobile doors are usually hemmed.

Piercing and cutting can also be performed in stamping presses. Progressive stamping is a combination of the above methods done with a set of dies in a row through which a strip of the material passes one step at a time.

5.4 冲压模具

冲压模具是在冷冲压加工中，将材料（金属或非金属）加工成零件（或半成品）的一种特殊工艺装备，称为冷冲压模具（俗称冷冲模）。冲压是在室温下，利用安装在压力机上的模具对材料施加压力，使其产生分离或塑性变形，从而获得所需零件的一种压力加工方法。

冲模的结构类型很多，为了研究方便，可以按冲模的不同特征进行分类。

◆ 按冲模完成的工序性质可分为：落料模、冲孔模、切断模、弯曲模、拉深模等几大类。

◆ 按工序的组合方式可分为：
(1) 单工序的简单模；
(2) 多工序的连续模；
(3) 复合模等。

简单模

在冲床的一次行程中只完成一道冲压工序的冲模称为简单模。其结构简单但效率低，适合于小批量、低精度的冲压件生产。

连续模

连续模（又称级进模、跳步模）是指压力机在一次行程中，依次在几个不同的位置上同时完成多道工序的冲模。

采用连续模冲压时，由于工件依次在不同的位置上逐步成形，为了控制工件的相对位置精度，应严格控制送料步距。连续模效率高且结构相对简单，适合于大批量、一般精度的冲压件生产。

复合模

复合模是指压力机在一次行程中，在同一中心位置上，同时完成几道工序的冲模。由于复合模是在同一中心位置上完成几道工序，因此它必须在同一中心位置上布置几套凸、凹模。

复合模效率高但结构复杂，适合于大批量、高精度的冲压件生产。

复合模

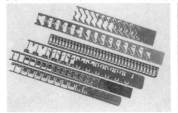
冲压件

汽车冲压件

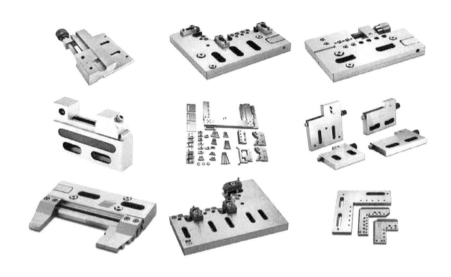

第三篇　切削加工与特种加工

　　汽车零件的表面形状千变万化，有不同的典型表面如外圆、内孔、平面、螺纹、花键和轮齿齿面等。这些典型表面都有一定的加工要求，大多数表面都需要经过专业加工来实现其制造过程。

　　金属切削加工是用刀具将金属毛坯逐层切削，使工件得到所需要的形状、尺寸和表面粗糙度的加工方法。

　　本篇主要介绍车削加工、铣削加工、磨削加工和电加工方法。

第六章 车削加工

6.1 概述

车床是主要用车刀对旋转的工件进行车削加工的机床。在车床上还可用钻头、扩孔钻、铰刀、丝锥、板牙和滚花工具等进行相应的加工。车床主要用于加工轴、盘、套和其他具有回转表面的工件，是机械制造和修配工厂中使用最广泛的一类机床。

下面我们就来聊聊被誉为"机器之母"——车床的前世今生。

车床雏形：树木车床

人类的祖先为了更方便地使用工具对木材进行加工，机床最早的原型——树木车床就这样诞生了。这就是车床的雏形，大概距今二千多年前吧。

13世纪，机床原型也在发展，就有了用脚踏板旋转曲轴并带动飞轮，再传动到主轴使其旋转的"脚踏车床"，也称为弹性杆棒车床，不过除了刀具是用金属外，操作原理还是跟原先一样的。

说到现代车床原型的诞生和普及，就不得不提"车床之父"——亨利·莫兹利，英国发明家。1797年，莫兹利制成了第一台螺纹切削车床，它带有丝杠和光杠，是采用了滑动刀架的现代车床，可车削不同螺距的螺纹。

莫兹利1797车床

此后，莫兹利又不断地对车床加以改进。1800年，他用坚实的铸铁床身代替了三角铁棒机架，用惰轮配合交换齿轮对，代替了更换不同螺距的丝杠来车削不同螺距的螺纹。

严格来说，车床并不是莫兹利发明的，他只是在前人的基础上，对车床进行了再创新，让它拥有了自动切削的功能。但莫兹利的再创造，让车床的应用得到了极大普及，车床得以真正意义上诞生。

莫兹利1800车床

进入19世纪后，由于各行业的发展，需要应用于不同类型的机床相继出现。1817年，罗伯茨发明了龙门车床。

机械加工、成型和安装工艺

罗伯茨龙门车床

1845 年，美国的菲奇发明了转塔车床。

1848 年，美国又出现了回轮车床。

1873 年，美国的斯潘塞制成一台单轴自动车床，不久他又制成三轴自动车床。

20 世纪初出现了由单独电机驱动的带有齿轮变速箱的车床。

这就是现代车床的原型，是具有划时代意义的刀架车床，对英国工业革命具有重要的意义。

第一次世界大战后，由于军火、汽车和其他机械工业的需要，各种高效自动车床和专门化车床迅速发展。为了提高小批量工件的生产率，20 世纪 40 年代末，带液压仿形装置的车床得到推广，与此同时，多刀车床也得到发展。

斯潘塞车床

带有齿轮变速箱的车床

我国车床的发展

由于历史原因，可以说，中国几乎完全错过了前三次的工业革命。中国车床真正发展是从建国后才开始的。新中国成立之初，包括机床在内的装备制造业可以说是一片空白。中国的第一批机床就是在这样"一穷二白"的条件下，经过工人们艰苦努力后成功下线的。

1949 年，新中国的第一台车床——六尺皮带车床，终于在沈阳第一机床厂诞生。

这批机床的问世及后来的批量生产，有力地支援了全国工业化建设，创造出了许多生产奇迹。

之后的近 30 年时间，由于国外技术封锁和国内基础条件限制，我国的数控技术并没有得到很好的发展。一直到改革开放后，我国才迎来了机床工业发展的春天，机床的种类及型号也逐渐丰富起来。

发展至今，立式、卧式、多轴、数控车、车铣复合等各类型车床应有尽有。如今，我国已连续多年成为世界第一大机床消费国和进口国，也是世界上能够生产所有门类机床的国家之一。

介绍完了历史，让我们扫一扫，了解车削的加工范围吧！

车削适于加工回转表面，还可打中心孔、钻孔、镗孔、车端面、切断、切槽、滚花、车螺纹、车锥体、车成形面(包括各种回转体表面)。

车削加工的工艺特点

1. 易于保证各加工面之间的位置精度。车削时工件作主运动，绕某一固定轴回转，各表面具有同一回转轴线。因此，各加工表面的位置精度容易控制和保证。

2. 切削过程比较平稳。车削的主运动为回转运动，避免了惯性力和冲击的影响。可进行高速切削或强力切削，有利于生产率的提高。

3. 刀具简单。车刀是机床刀具中最简单的一种，制造、刃磨和安装都比较方便。

History

The lathe is an ancient tool, with the first known representation dating to the 3rd century BC in ancient Egypt. Clear evidence of turned artifacts have been found from the 6th century BC: fragments of a wooden bowl in an Etruscan tomb in Northern Italy as well as two flat wooden dishes with decorative turned rims from modern Turkey. More tenuous evidence existed from a Mycanean Greek site as far back as the 13th or 14th century BC. There was also evidence of use in Assyria and India. The lathe was very important to the Industrial Revolution. It is known as the *mother of machine tools*, as it was the first machine tool that led to the invention of other machine tools. The ancient Chinese dating to the Warring States era also used rotary lathes to sharpen tools and weapons on an industrial scale.

The origin of turning dates to around 1300 BC when the Ancient Egyptians first developed a two-person lathe. One person would turn the wood work piece with a rope while the other used a sharp tool to cut shapes in the wood. Ancient Rome improved the Egyptian design with the addition of a turning bow. In the Middle Ages a pedal replaced hand-operated turning, allowing a single person to rotate the piece while working with both hands. The pedal was usually connected to a pole, often a straight-grained sapling. The system today is called the "spring pole" lathe. Spring pole lathes were in common use into the early 20th century.

An important early lathe in the UK was the horizontal boring machine that was installed in 1772 in the Royal Arsenal in Woolwich. It was horse-powered and allowed for the production of much more accurate and stronger cannon used with success in the American Revolutionary War in the late 18th century. One of the key characteristics of this machine was that the workpiece was turning as opposed to the tool, making it technically a lathe (see attached drawing). Henry Maudslay who later developed many improvements to the lathe worked at the Royal Arsenal from 1783 being exposed to this machine in the Verbruggen workshop.

During the Industrial Revolution, mechanized power generated by water wheels or steam engines was transmitted to the lathe via line shafting, allowing faster and easier work. Metalworking lathes evolved into heavier machines with thicker, more rigid parts. Between the late 19th and mid-20th centuries, individual electric motors at each lathe replaced line shafting as the power source. Beginning in the 1950s, servomechanisms were applied to the control of lathes and other machine tools via numerical control, which often was coupled with computers to yield computerized numerical control (CNC). Today manually controlled and CNC lathes coexist in the manufacturing industries.

Turning

Turning is a machining process in which a cutting tool, typically a non-rotary tool bit, describes a helix toolpath by moving more or less linearly while the workpiece rotates. The tool's axes of movement may be literally a straight line, or they may be along some set of curves or angles, but they are essentially linear (in the non mathematical sense). Usually the term "turning" is reserved for the generation of external surfaces by this cutting action, whereas this same essential cutting action when applied to internal surfaces (that is, holes, of one kind or another) is called "boring". Thus the phrase "turning and boring" categorizes the larger family of (essentially similar) processes known as lathing. The cutting of faces on the workpiece (that is, surfaces perpendicular to its rotating axis), whether with a turning or boring tool, is called "facing", and may be lumped into either category as a subset.

Turning can be done manually, in a traditional form of lathe, which frequently requires continuous supervision by the operator, or by using an automated lathe which does not. Today the most common type of such automation is computer numerical control, better known as CNC. (CNC is also commonly used with many other types of machining besides turning.)

When turning, a piece of relatively rigid material (such as wood, metal, plastic, or stone) is rotated and a cutting tool is traversed along 1, 2, or 3 axes of motion to produce precise diameters and depths. Turning can be either on the outside of the cylinder or on the inside (also known as boring) to produce tubular components to various geometries. Although now quite rare, early lathes could even be used to produce complex geometric figures, even the platonic solids; although since the advent of CNC it has become unusual to use non-computerized toolpath control for this purpose.

The turning processes are typically carried out on a lathe, considered to be the oldest machine tools, and can be of four different types such as straight turning, taper turning, pro filing or external grooving. Those types of turning processes can produce various shapes of materials such as straight, conical, curved, or grooved workpiece. In general, turning uses simple single-point cuttingtools. Each group of workpiece materials has an optimum set of tools angles which have been developed through the years.

The bits of waste metal from turning operations are known as chips (North America), or swarf (Britain). They may be known as turnings somewhere.

Turning operations

Turning specific operations include:

1. Turning

This operation is one of the most basic machining processes. That is, the part is rotated while a single point cutting tool is moved parallel to the axis of rotation. Turning can be done on the external surface of the part as well as internally (boring). The starting material is generally a workpiece generated by other processes such as casting, forging, extrusion, or drawing.

(1) Tapered turning

From the compound slide

From taper turning attachment

Using a hydraulic copy attachment

Using a C.N.C. lathe

Using a form tool

By the offsetting of the tailstock - this method more suited for shallow tapers.

(2) Spherical generation

The proper expression for making or turning a shape is to generate as in to generate a form around a fixed axis of revolution.

a) using hydraulic copy attachment.

b) C.N.C. (computer numerical controlled) lathe.

c) using a form tool (a rough and ready method).

d) using bed jig (need drawing to explain).

(3) Hard turning

Hard turning is a turning done on materials with a Rockwell C hardness greater than 45. It is typically performed after the workpiece is heat treated.

The process is intended to replace or limit traditional grinding operations. Hard turning, when applied for purely stock removal purposes, competes favorably with rough grinding. However, when it is applied for finishing where form and dimension are critical, grinding is superior. Grinding produces higher dimensional accuracy of roundness and cylindricity. In addition, polished surface finishes of $Rz=0.3-0.8z$ cannot be achieved with hard turning alone. Hard turning is appropriate for parts requiring roundness accuracy of 0.5-12 micrometres, and/or surface roughness of Rz 0.8 – 7.0 micrometres. It is used for gears, injection pump components, hydraulic components, among other applications.

2. Facing

Facing in the context of turning work involves moving the cutting tool at right angles to the axis of rotation of the rotating workpiece. This can be performed by the operation of the cross-slide, if one is fitted, as distinct from the longitudinal feed (turning). It is frequently the first operation performed in the production of the workpiece, and often the last-hence the phrase "ending up".

3. Parting

This process, also called parting off or cutoff, is used to create deep grooves which will remove a completed or part-complete component from its parent stock.

4. Grooving

Grooving is like parting, except that grooves are cut to a specific depth instead of severing a completed/part-complete component from the stock. Grooving can be performed on internal and external surfaces, as well as on the face of the part (face grooving or trepanning).

Non-specific operations include:

(1) Boring

Enlarging or smoothing an existing hole created by drilling, moulding etc.i.e. the machining of internal cylindrical forms (generating) a) by mounting workpiece to the spindle via a chuck or faceplate b) by mounting workpiece onto the cross slide and placing cutting tool into the chuck. This work is suitable for castings that are too awkward to mount in the face plate. On long bed lathes large workpiece can be bolted to a fixture on the bed and a shaft passed between two lugs on the workpiece and these lugs can be bored out to size. A limited application but one that is available to the skilled turner/machinist.

(2) Drilling

It is used to remove material from the inside of a workpiece. This process utilizes standard drill bits held stationary in the tail stock or tool turret of the lathe. The process can be done by separately available drilling machines.

(3) Knurling

The cutting of a serrated pattern onto the surface of a part to use as a hand grip using a special purpose knurling tool.

(4) Reaming

The sizing operation that removes a small amount of metal from a hole already drilled. It is done for making internal holes of very accurate diameters. For example, a 6mm hole is made by drilling with 5.98 mm drill bit and then reamed to accurate dimensions.

(5) Threading

Both standard and non-standard screw threads can be turned on a lathe using an appropriate cutting tool. (Usually having a 60, or 55° nose angle) Either externally, or within a bore. Generally referred to as single-point threading.

Tapping of threaded nuts and holes:

a) using hand taps and tailstock centre

b) using a tapping device with a slipping clutch to reduce risk of breakage of the tap.

Threading operations include:

a) All types of external and internal thread forms using a single point tool also taper threads, double start threads, multi start threads, worms as used in worm wheel reduction boxes, leadscrew with single or multistart threads.

b) By the use of threading boxes fitted with 4 form tools, up to 2" diameter threads but it is possible to find larger boxes than this.

(6) Polygonal turning

Non-circular forms are machined without interrupting the rotation of the raw material.

6.2　车床

卧式车床

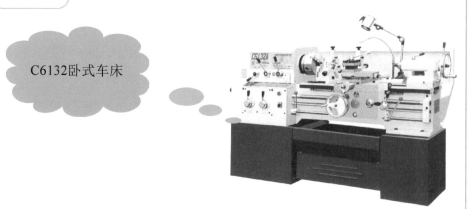

C6132卧式车床

卧式车床的主轴是水平的,主要用车刀对旋转的工件进行车削加工。在车床上还可用钻头、扩孔钻、铰刀、丝锥、板牙和滚花工具等进行相应的加工。

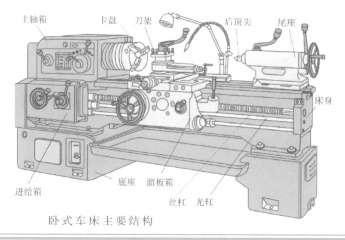

卧式车床主要结构

卧式车床的特点：

◆车床的床身、床脚、油盘等采用整体铸造结构，刚性高，抗震性好，符合高速切削机床的特点。

◆床头箱采用三支承结构，三支承均为圆锥滚子轴承，主轴调节方便，回转精度高，精度保持性好。

◆进给箱设有公英制螺纹转换机构，螺纹种类的选择转换方便、可靠。

◆溜板箱内设有锥形离合器安全装置，可防止自动走刀过载后的机件损坏。

◆车床纵向设有四工位自动进给机械碰停装置，可通过调节碰停杆上的凸轮纵向位置，设定工件加工所需长度，实现零件的纵向定尺寸加工。

◆尾座设有变速装置，可满足钻孔、铰孔的需要。

◆车床润滑系统设计合理可靠，床头箱、进给箱、溜板箱均采用体内飞溅润滑，并增设线泵、柱塞泵对特殊部位进行自动强制润滑。

C61200重型卧式车床

立式车床

立式车床的主轴是垂直的，并有一安装工件的圆形工作台。由于工作台处于水平位置，工件的找正和夹紧比较方便，且工件及工作台的重力由床身导轨或推力轴承承受，主轴不产生弯曲。因此立式车床适用于加工较大的盘类及大而短的套类零件。立式车床上的垂直刀架可沿横梁导轨和刀架座导轨移动，作横向或纵向进给。刀架座可偏转一定角度作斜向进给。侧刀架可沿立柱导轨上下移动，也可沿刀架滑座左右运动，实现纵向或横向进给。

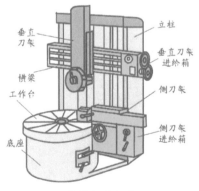

单柱立式车床　　　　　　　　　单柱立式车床

双柱立式车床

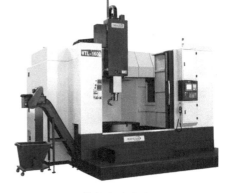
数控立式车床

立式车床适用于加工直径大、轴向尺寸相对较小、高度与直径比在0.32～0.8之间、形状复杂的大型和重型工件，如机架、体壳类零件等。可以进行内外圆柱面、圆锥面、端面、沟槽、切断以及钻、扩、镗和铰孔等加工，也可借助于附加装置车螺纹、车球面、仿形铣削和磨削等。立式车床通常用于单件小批生产。立式车床是气轮机、水轮机、矿山冶金等重型机械制造不可缺少的设备，一般机械制造使用也很普遍。根据立柱的数目可分为单柱立式车床和双柱立式车床。单柱立式车床是轻型立式车床，刚度没有双柱立式车床好。根据横梁是否固定可分为横梁固定式和横梁升降式，后者加工范围较大。

落地车床

落地车床直接将主轴箱安装在地基上。落地车床主要用于车削大型工件端面，适用于直径大、长度短、质量较小的盘、环、薄壁筒形工件，车内外圆柱面、圆锥面、端面、切槽、切断等。加辅助装置时也可车螺纹、磨削、仿形车等。

落地车床的底座导轨采用矩形结构，跨距大、刚性好，适用于低速重载切削。操纵杆安装在前床腿位置，操作方便，外观协调。落地车床分为连体落地车床、分体落地车床和重型落地车床。

落地车床多用于轻工、化工、造纸、水泥、冶金、船舶、港口和矿山机械等行业。

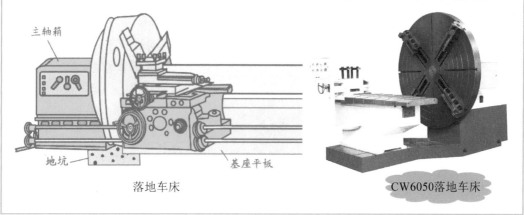

落地车床　　　　　　　　　　CW6050落地车床

转塔车床

转塔车床（又称六角车床）有六工位转塔刀架，转塔刀架轴线垂直于机床主轴，可沿导轨作纵向运动，转塔车床多有前、后刀架可作纵、横移动。转塔刀架各刀具均按加工顺序预调好，切削一次后，刀架退回并转位，再用另一把刀切削，故可在一次装夹中完成较复杂的加工。用可调挡块控制刀具行程终点位置。转塔刀架和横刀架适当调整可以联合车削。

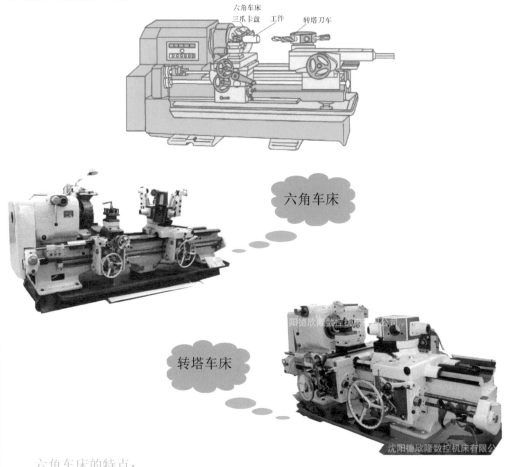

六角车床的特点：
- ◆ 能实现多刀切削、工序集中。加工效率比一般卧式车床高2～3倍。
- ◆ 可设撞停装置。自动控制工件尺寸，保证成批工件尺寸的一致性。
- ◆ 机床刚性好。可满足多刀架复合切削和多刀刃切削的要求。

仿形车床

仿形车床是由仿形刀架按样件或样板表面，作纵、横向随动运动，车刀便复制出相应形状的被加工零件。仿形车床适用于大批大量生产的圆柱形、圆锥形、台阶形及其他成形旋转曲面的轴、套、盘、环类工件的车削加工。所有动作均可由液压传动实现，响应速度快，无级调速，结构简单紧凑。液压仿形车床电气控制可采用可编程序控制器（PLC）控制。用仿形车床加工形状复杂的工件时，其生产率比卧式车床高 15 倍左右，工件形状越复杂、批量越大，则效率越高。

仿形车床是指能仿照样板或样件的形状尺寸，自动完成工件的加工循环的数控车床，适用于形状较复杂的工件的小批和成批生产，生产率比普通车床高 10～15 倍。有多刀架、多轴、卡盘式、立式等类型。

右图为车床的液压随动作用式仿形装置。随动作用式仿形原理是把样板给仿形触头的位移信号转换成电信号（电压）或液压信号（压力差），经功率放大后驱动机床执行部件。驱动元件可以是直流电机、油缸或液压马达等。采用这种控制方式的样板和触头承受压力较小。随动阀接受样板和触头的位移信号后，通过液压油路作用于油缸活塞，使车刀执行仿形运动。

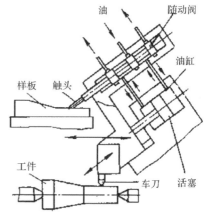

C7250 液压多刀仿形车床

介绍了这么多种车床，那车床的主要部件都有哪些呢？它们的作用又是什么呢？

> 三箱：主轴箱、进给箱、溜板箱
> 两架：刀架、尾架
> 一身：床身

我们以普通卧式车床为例进行介绍。

◆主轴箱：又称为床头箱、变速箱，是位于车床左上方的箱体。

◇作用：

(1) 支承主轴，将电动机的动力传给主轴，使主轴带动工件旋转。

(2) 可以通过变速齿轮机构，使主轴获得多种不同转速。

(3) 主轴为空心结构，前部外锥面安装附件（卡盘），是用来夹持工件的。当用三爪夹持工件时，可以实现自动定心。前部内锥面用来安装顶尖。细长孔可以穿入长棒料。

◇主轴转速：是以每分钟多少转来计算的，单位为转/分。

◆进给箱：又称走刀箱，其内装进给运动的变速机构，用来传递进给运动。

◇作用：改变手柄位置，可以得到不同的进给速度，并可将主轴的运动分别传递给光杠和丝杠。传给光杠可以得到各种不同的走刀量；传给丝杠可以车削各种不同螺距的螺纹。

◇进给量：是指工件每旋转一周，车刀沿走刀方向移动的距离，单位为毫米/转。

◆溜板箱：又称拖板箱，它是车床进给运动的操纵箱。

◇作用：它能够改变运动，可以将光杠传来的旋转运动变为车刀需要的纵向或横向的直线运动；也可操纵对开螺母使刀架由丝杠直接带动车削螺纹。

◆刀架：由多层结构组成，它分成纵溜板、横溜板、小溜板、转盘、方刀架等五个部分。

◇纵溜板：又称大拖板、大刀架，是控制车刀纵向车削及控制工件长度尺寸用的。它与溜板箱连接，带动车刀沿床身导轨作纵向移动。它上面有横向导轨。

◇横溜板：又称中溜板，是横向车削工件及控制吃刀深度用的。它可在纵溜板上面的导轨上作横向移动。

◇转盘：它与横溜板用螺钉紧固。松开螺钉，便可在水平面内偏转任意角度。其上面有小溜板的导轨。

◇小溜板：是控制长度方向的微量切削的。它可沿转盘上面的导轨作短距离的纵向移动。将转盘偏转若干角度后，小溜板作斜向进给，可以车削圆锥体。

◇方刀架：它固定在小溜板上，可装四把车刀。松开锁紧手柄，转动小刀架，可以把所需的车刀转到工作位置上。工作时，必须把手柄扳紧。

◆尾架：安装于床身导轨上，其位置可沿床身导轨调节。

◇作用：它主要用来配合主轴支持工件或者是装夹钻头及孔加工刀具，它还可以用来套丝以及装夹加工需要的专用刀具。将尾架体偏移还可用来车削锥体。

◆床身：主要用来支承和连接机床的各个部件，并且保持各部件的相对正确位置。

床身上面有两组精度很高的导轨:一组供刀架移动用;一组供尾架移动用。床身由床脚支承固定在地基上。

文字太多,就有点理解困难了!

所以才需要扫一扫呀!在此了解车床的主要结构、各执行机构及运动方式。

Category
    Horizontal lathe
    Floor-type horizontal lathe
    Turret lathe
    Vertical lathe
    Copying lathe

Horizontal lathe

    Horizontal lathe is mainly manufacture for turning rotating workpieces. Drills, reaming drills, reamers, taps, plate teeth and knurling tools can also be used for corresponding machining.

Horizontal lathe features:

◆ The lathe bed, foot and oil pan adopt integral casting structure, which has high rigidity and good seismic resistance, and accords with the characteristics of high-speed machine tools.

◆ The headbox adopts three supporting structures, all of them are tapered roller bearings. The spindle is easy to adjust, the rotary accuracy is high and the accuracy is maintained well.

◆ Feed box is equipped with a Metric-British threading conversion mechanism. The selection and conversion of threading types are convenient and reliable.

◆ The safety device of conical clutch is installed in the slide box, which can prevent the damage of the machine parts after overloading of the automatic cutter.

◆ The lathe is equipped with four-position automatic feeding machine collision-stop device in the longitudinal direction. The longitudinal dimension of the parts can be achieved by adjusting the cam longitudinal position on the collision-stop bar and setting the required length of the workpiece.

◆ The tailstock is equipped with a gearshift device, which can meet the needs of drilling and reaming.

◆ The design of lathe lubrication system is reasonable and reliable. Splash lubrication is used in headstock, feed box and slide box. Line pump and plunger pump are added to automatically force lubrication for special parts.

Vertical lathe

The spindle of the vertical lathe is vertical and has a circular worktable for installing workpieces. It is more convenient to align workpiece and clamp for the worktable is in the horizontal position, and the gravity of the workpiece and worktable is borne by the bed guide or thrust bearing, the spindle does not bend. Therefore, the vertical lathe is suitable for processing large discs and large and short sets of parts. The vertical tool holder on the vertical lathe can move along the cross beam guide rail and the tool holder guide rail for transverse or longitudinal feed. The tool holder can be deflected to a certain angle for oblique feed. The side tool holder can move up and down along the column guide rail, or move left and right along the sliding seat of the tool holder to realize vertical or horizontal feeding.

Landing lathe

The landing lathe directly installs the spindle box on the foundation. The floor lathe is used mainly for turning the end face of large workpiece. It is suitable for disc, ring and thin-walled cylindrical workpiece with large diameter, short length and small mass. The inside and outside cylindrical surface, conical surface, end face, grooving, cutting and so on. Adding auxiliary devices can also turn threads, grinding, copying cars, etc.

The floor lathe can be divided into two types: base lathe and elevation lathe. The base type spindle box landed on the ground. There was a pit between the spindle box and the base plate. The tool holder can be placed in many positions vertically and horizontally on the base plate. The knife holder moves vertically and horizontally on the big knife holder, and the knife holder moves vertically and horizontally on the base plate, thus eliminating the bed.

Landing lathes are mostly used in light industry, chemical industry, paper making, cement, metallurgy, ship, port and mining machinery industries.

Turret lathe

Hexagonal lathe (also known as turret lathe) has six-position turret tool holder. The axis of turret tool holder is perpendicular to the main axis of the machine tool. It can move along the guide rail longitudinally. Turret lathe has front and rear tool holders which can move longitudinally and horizontally. Each tool of turret tool holder is pre-adjusted according to the processing sequence. After cutting once, the tool holder is retracted and rotated, and then cut with another tool. Therefore, complex processing can be completed in one clamp. Adjustable block is used to control the end position of tool stroke. The turret tool holder and the transverse tool holder can be combined cutting by proper adjustment.

Hexagonal lathe features:

◆ It can realize multi-tool cutting and process concentration. The processing efficiency is 2-3 times higher than that of the general horizontal lathe.

◆ Collision stop device can be installed. Automatic control of workpiece size to ensure the consistency of workpiece size in batches.

◆ Good rigidity of machine tools. It can meet the requirements of multi-tool holder compound cutting and multi-tool edge cutting.

Copying lathe

The copying lathe is made up of the copying tool holder according to the sample or the surface of the template, which makes the longitudinal and lateral follow-up movement, and the turning tool can copy the corresponding shape of the processed parts. It is used for turning cylindrical, conical, step-shaped and other shaping revolving surface of axes, sleeves, discs and rings in mass production. All actions can be realized by hydraulic transmission, with fast response, stepless speed regulation and simple and compact structure. The electric control of hydraulic copying lathe can be controlled by programmable logic controller (PLC). The productivity of the workpiece with complex shape is about 15 times higher than that of the horizontal lathe. The more complex the shape of the workpiece and the larger the batch, the higher the efficiency.

Profile-copying lathe refers to a CNC lathe that can automatically complete the processing cycle of workpiece by imitating the shape and size of the template or sample. It is suitable for small batch and batch production of workpiece with complex shape, and its productivity is 10-15 times higher than that of ordinary lathe. There are many types of tool holder, multi-axis, chuck type, vertical type, etc.

Construction of lathe

The design of lathes can vary greatly depending on the intended application; however, basic features are common to most types. These machines consist of (at the least) a headstock, bed, carriage, and tailstock. Better machines are solidly constructed with broad bearing surfaces (*slide-ways*) for stability, and manufactured with great precision. This helps ensure the components manufactured on the machines can meet the required tolerances and repeatability.

Headstock

The headstock houses the main spindle, speed change mechanism , and change gears. The headstock is required to be made as robust as possible due to the cutting forces involved, which can distort a lightly built housing, and induce harmonic vibrations that will transfer through to the workpiece, reducing the quality of the finished workpiece.

The main spindle is generally hollow to allow long bars to extend through to the work area. This reduces preparation and waste of material. The spindle runs in precision bearings and is fitted with some means of attaching workholding devices such as chucks or faceplates. This end of the spindle usually also has an included taper, frequently a Morse taper, to allow the insertion of hollow tubular (Morse standard) tapers to reduce the size of the tapered hole, and permit use of centers. On older machines ('50s) the spindle was directly driven by a flat belt pulley with lower speeds available by manipulating the bull gear. Later machines use a gear box driven by a dedicated electric motor. A fully "geared head" allows the operator to select suitable speeds entirely through the gearbox.

Beds

The bed is a robust base that connects to the headstock and permits the carriage and tailstock to be moved parallel with the axis of the spindle. This is facilitated by hardened and ground bedways which restrain the carriage and tailstock in a set track. The carriage travels by means of a rack and pinion system. The leadscrew of accurate pitch, drives the carriage holding the cutting tool via a gearbox driven from the headstock.

Types of beds include inverted "V" beds, flat beds, and combination "V" and flat beds. "V" and combination beds are used for precision and light duty work, while flat beds are used for heavy duty work.

When a lathe is installed, the first step is to *level* it, which refers to making sure the bed is not twisted or bowed. There is no need to make the machine exactly horizontal, but it must be entirely untwisted to achieve accurate cutting geometry. A precision level is a useful tool for identifying and removing any twist. It is advisable also to use such a level along the bed to detect bending, in the case of a lathe with more than four mounting points. In both instances the level is used as a comparator rather than an absolute reference.

Feed and lead screws

The feed screw is a long driveshaft that allows a series of gears to drive the carriage mechanisms. These gears are located in the apron of the carriage. Both the feedscrew and leadscrew are driven by either the change gears (on the quadrant) or an intermediate gearbox known as a quick change gearbox or Norton gearbox. These intermediate gears allow the correct ratio and direction to be set for cutting threads or worm gears. Tumbler gears are provided between the spindle and gear train along with a quadrant plate that enables a gear train of the correct ratio and direction to be introduced. This provides a constant relationship between the number of turns the spindle makes, to the number of turns the leadscrew makes. This ratio allows screw threads to be cut on the workpiece without the aid of a die.

Some lathes have only one leadscrew that serves all carriage-moving purposes. For screw cutting, a half nut is engaged to be driven by the leadscrew's thread; and for general power feed, a key engages with a keyway cut into the leadscrew to drive a pinion along a rack that is mounted along the lathe bed.

The leadscrew will be manufactured to either imperial or metric standards and will require a conversion ratio to be introduced to create thread forms from a different family. To accurately convert from one thread form to the other requires a 127-tooth gear, or on lathes not large enough to mount one, an approximation may be used. Multiples of 3 and 7 giving a ratio of 63∶1 can be used to cut fairly loose threads. This conversion ratio is often built into the *quick change gearboxes.*

Carriage

In its simplest form the carriage holds the tool bit and moves it longitudinally (turning) or perpendicularly (facing) under the control of the operator. The operator moves the carriage manually via the *handwheel* or automatically by engaging the feed shaft with the carriage feed mechanism. This provides some relief for the operator as the movement of the carriage becomes power assisted. The handwheels, 5aon the carriage and its related slides are usually calibrated, both for ease of use and to assist in making reproducible cuts. Calibration marks will measure either the distance from center (radius), or the diameter of the workpiece, for example, on a diameter machine where calibration marks are in thousandths of an inch, the radial handwheel dial will read 0.0005 inches of radius per division, or 0.001 inches of diameter. The carriage typically comprises a top casting, known as the saddle, and a side casting, known as the apron.

1. Cross-slide

The cross-slide rides on the carriage and has a feedscrew that travels at right angles to the main spindle axis. This permits facing operations to be performed, and the depth of cut to be adjusted. This feedscrew can be engaged, through a gear train, to the feed shaft (mentioned previously) to provide automated "power feed" movement to the cross-slide. On most lathes, only one direction can be engaged at a time as an interlock mechanism will shut out the second gear train.

2. Compound rest

The compound rest (or top slide) is usually where the tool post is mounted. It provides a smaller amount of movement (less than the cross-slide) along its axis via another feedscrew. The compound rest axis can be adjusted independently of the carriage or cross-slide. It is used for turning tapers, to control depth of cut when screwcutting or precision facing, or to obtain finer feeds (under manual control) than the feed shaft permits. Usually, the compound rest has a protractor marked in its base, enabling the operator to adjust its axis to precise angles.

The slide rest (as the earliest forms of carriage were known) can be traced to the fifteenth century. In 1718 the tool-supporting slide rest with a set of gears was introduced by a Russian inventor Andrey Nartov and had limited usage in the Russian industry. In the eighteenth century the slide rest was also used on French ornamental turning lathes. The suite of gun boring mills at the Royal Arsenal, Woolwich, in the 1780s by the Verbruggan family also had slide rests. The story has long circulated that Henry Maudslay invented it, but he did not (and never claimed so). The legend that Maudslay invented the slide rest originated with James Nasmyth, who wrote ambiguously about it in his Remarks on the Introduction of the Slide Principle, 1841; later writers misunderstood, and propagated the error. However, Maudslay did help to disseminate the idea widely. It is highly probable that he saw it when he was working at the Arsenal as a boy. In 1794, whilst he was working for Joseph Bramah, he made one, and when he had his own workshop used it extensively in the lathes he made and sold there. Coupled with the network of engineers he trained, this ensured the slide rest became widely known and copied by other lathe makers, and so diffused throughout British engineering workshops. A practical and versatile screw-cutting lathe incorporating the trio of leadscrew, change gears, and slide rest was Maudslay's most important achievement.

The first fully documented, all-metal slide rest lathe was invented by Jacques de Vaucanson around 1751. It was described in the Encyclopédie a long time before Maudslay invented and perfected his version. It is likely that Maudslay was not aware of Vaucanson's work, since his first versions of the slide rest had many errors that were not present in the Vaucanson lathe.

3. Toolpost

The tool bit is mounted in the toolpost which may be of the American lantern style, traditional four-sided square style, or a quick-change style such as the multifix arrangement pictured. The advantage of a quick change set-up is to allow an unlimited number of tools to be used (up to the number of holders available) rather than being limited to one tool with the lantern style, or to four tools with the four-sided type. Interchangeable tool holders allow all tools to be preset to a center height that does not change, even if the holder is removed from the machine.

Tailstock

The tailstock is a tool (drill), and centre mount, opposite the headstock. The spindle does not rotate but does travel longitudinally under the action of a leadscrew and handwheel. The spindle includes a taper to hold drill bits, centers and other tooling. The tailstock can be positioned along the bed and clamped in position as dictated by the work piece. There is also provision to offset the tailstock from the spindles axis, this is useful for turning small tapers, and when re-aligning the tailstock to the axis of the bed.

The image shows a reduction gear box between the handwheel and spindle, where large drills may necessitate the extra leverage. The tool bit is normally made of HSS, cobalt steel or carbide.

Steady, follower and other rests

Long workpieces often need to be supported in the middle, as cutting tools can push (bend) the workpiece away from where the centers can support them, because cutting metal produces tremendous forces that tend to vibrate or even bend the workpiece. This extra support can be provided by a steady rest (also called a steady, a fixed steady, a center rest, or sometimes, confusingly, a center). It stands stationary from a rigid mounting on the bed, and it supports the workpiece at the rest's center, typically with three contact points 120° apart. A follower rest (also called a follower or a travelling steady) is similar, but it is mounted to the carriage rather than the bed, which means that as the tool bit moves, the follower rest "follows along" (because they are both rigidly connected to the same moving carriage).

Follower rests can provide support that directly counteracts the springing force of the tool bit, right at the region of the workpiece being cut at any moment. In this respect they are analogous to a box tool. Any rest transfers some workpiece geometry errors from base (bearing surface) to processing surface. It depends on the rest design. For minimum transfer rate correcting rests are used. Rest rollers typically cause some additional geometry errors on the processing surface.

6.3 车床常用附件

车削工件时,通常总是先把工件装夹在车床的卡盘或夹具上,经过校正,然后进行加工。车床夹具的主要作用是确定工件在车床上的正确位置,并可靠地夹紧工件。常用的车床夹具有以下几类。

◆ 通用夹具或附件:如三爪卡盘、四爪卡盘、花盘、各种形式的顶针、中心架和跟刀架等。

◆ 可调夹具:如成组夹具、组合夹具等。

◆ 专用夹具:专门为满足某个零件的某道工序而设计的夹具。根据工件的特点,可利用不同的附件进行不同方法的装夹。在各种批量的生产中,正确地选择和使用夹具,以保证质量,提高生产效率,减轻劳动强度密切相关。在实际中,主要使用通用夹具。

> 下面仅介绍常用的通用夹具和附件的装夹方法。

三爪卡盘

三爪卡盘是车床上最常用的通用夹具,一般由专业厂家生产,作为车床附件配套供应。三爪卡盘的特点是装夹工件能自动定心,装夹方便,可省去许多校正工作,适用于装夹中心型短棒料或圆盘类等工件。但其定心准确度不太高(0.05～0.15 mm)。

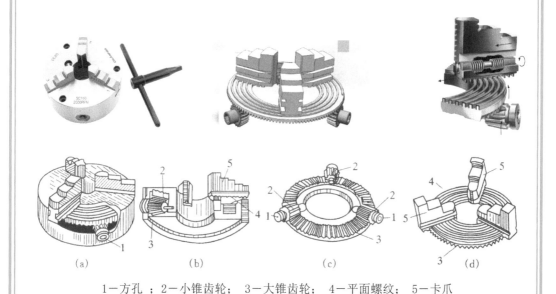

1—方孔;2—小锥齿轮;3—大锥齿轮;4—平面螺纹;5—卡爪

四爪卡盘

四爪卡盘也是常用的通用夹具，由专业厂家生产，以车床附件成套供应。四爪卡盘的特点是夹紧力大，用途广泛。虽不能自动定心，但通过校正后，安装精度较高，可达 0.01 mm。

四爪卡盘适于装夹圆形、方形、长方形、椭圆或其他不规则形状的工件。在圆盘上车偏心孔也常用四爪卡盘。如果把四个卡爪各自掉头安装到卡盘上，可起到"反爪"作用，即可安装较大的工件。使用四爪卡盘装夹工件，调整时间长，对工人技术水平要求高，故四爪卡盘在单件小批量生产中常用来装夹不规则的零件。

中心架及跟刀架

车削长度为直径 10 倍以上的细长轴时，由于工件本身的刚性不足，在工件重量和切削力的作用下，工件会产生弯曲变形，影响加工精度。为了防止工件的弯曲，这时应采用附加辅助支承——中心架或跟刀架。

中心架被压紧在床身上，以其三个互成 120°的支承爪把工件支承住进行加工。因为中心架是被压紧在床身上的，所以大拖板不能超过它，因此加工长工件时，需先加工一端，然后掉头安装再加工另一端。

通常使用中心架的情况是：
(1) 车削不能通过机床主轴孔的大直径长杆件端面。
(2) 在长杆件端部进行钻孔、镗孔或攻丝。
(3) 一般多用于加工阶梯轴。

跟刀架是被固定在车床大拖板上使用的。它与刀具一起移动，将两个支承点调整到恰好跟在刀具后面移动，这样就使带衬垫的支承点支承在刚刚加工好的表面上。跟刀架主要用于那些在卡盘上掉头安装不方便的细长轴加工。

应用跟刀架或中心架时，工件被支承部位应是加工过的外圆表面，并要加机油润滑。工件的转速不能过高，以免工件与支承爪之间摩擦过热而烧坏或磨损支承爪。

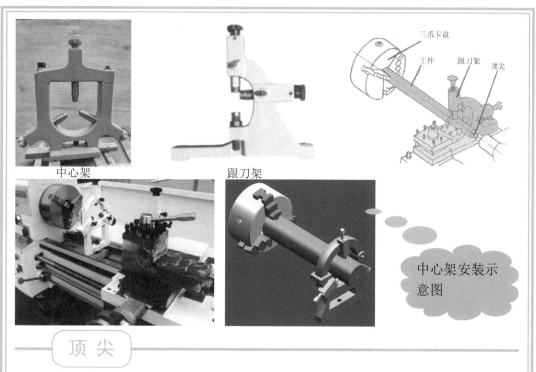

中心架　　　　跟刀架　　　　　　　　　中心架安装示意图

顶　尖

顶尖适合于安装长轴类工件进行低速精加工和半精加工。顶尖分前顶尖和后顶尖两类。

◆前顶尖：插在主轴锥孔内与主轴一起旋转的顶尖称作前顶尖。前顶尖随工件一起转动，与中心孔无相对运动，不发生摩擦。

◆后顶尖：插入车床尾座套筒内的顶尖称为后顶尖，后顶尖又分为固定顶尖和回转顶尖两种。

以两顶尖来装夹工件，多用于工件在加工过程中需多次装夹，而又要求有同一定位基准的情况。为了保护已加工表面，需在工件和紧固螺钉间放一软垫片，如铜片等。

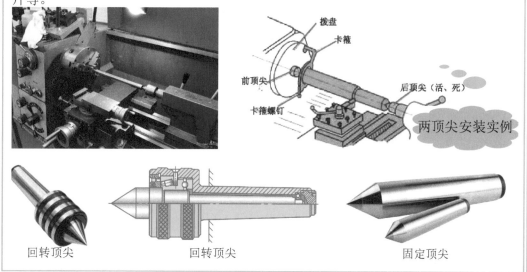

两顶尖安装实例

回转顶尖　　　　回转顶尖　　　　固定顶尖

芯轴

芯轴安装是以工件内孔为基准保证零件加工的位置精度，中小型的套、带轮等零件，一般可用芯轴安装。芯轴制造容易，使用方便，因此在工厂应用很广泛。有些形状或同轴度要求较高的盘套类零件，需用芯轴安装进行加工，这时应先加工孔，然后以孔定位，安装在芯轴上加工外圆。根据工件的形状尺寸及精度要求和加工数量的不同，应采用不同结构的芯轴。

◆ 小锥度芯轴：当工件长度大于工件孔径时，可采用稍带有锥度的小锥度芯轴。
◆ 圆柱芯轴：当工件长度比孔径小时，则应做成带螺母压紧的圆柱芯轴。
◆ 胀力芯轴：适用于中小型工件的安装。它是通过调整锥形螺钉使芯轴一端作微量的径向扩张，将工件胀紧。

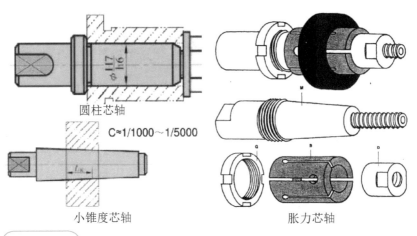

花盘

花盘装夹常用于装夹形状复杂的工件。在花盘上装夹工件时，找正比较费时。同时，要用平衡铁平衡工件和弯板等，以防止旋转时发生振动。

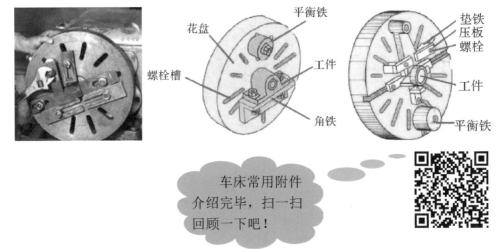

Accessory

When turning a workpiece, the workpiece is usually clamped on the chuck or fixture of the lathe, and to be adjusted, then processed. The main function of lathe fixture is to determine the correct position of the workpiece on the lathe and clamp the workpiece reliably. The commonly used lathe fixtures are:

◆ General fixture or accessories: such as three-jaw chuck, four-jaw chuck, disc chuck, various forms of thimble, steady rest and follow rest, etc.

◆ Adjustable fixture: such as group fixture, modular fixture, etc.

◆ Special Fixture: It is designed for a special part or a process. According to the characteristics of the workpiece, which can be used for different clamping methods. In all kinds of production, the correct selection and use of fixtures are closely related to ensuring quality, improving production efficiency and reducing labor intensity. In practice, the general fixture is mainly used.

Three-jaw chuck

The three-jaw chuck is the most commonly used fixture on lathe, which is generally produced by professional manufacturers and supplied as a complete set of lathe accessories. Three-jaw chuck is characterized by automatic centering of clamping workpiece, convenient clamping, and can save a lot of calibration work. It is suitable for clamping workpieces such as short central bar or disc. But the centering accuracy is not very high (0.05-0.15 mm).

Four-jaw chuck

Four-jaw chuck is also the most commonly used fixture. It is also manufactured by professional manufacturers and supplied with lathe accessories. Four-jaw chuck is characterized by large clamping force and wide application. Although it can not be automatically centered, but the installation accuracy is high after being adjusted, the accuracy can be reached 0.01 mm.

Four-jaw chuck is suitable for clamping round, square, rectangular, elliptical or other irregular shaped workpieces. Four-jaw chucks are also commonly used for machining of eccentric holes on discs. If the reversed four claws are installed on the chuck head to play the role of "anti-claw", the larger workpiece can be installed. Four-jaw chuck is often used to clamp irregular parts in small batch production because of its long adjustment time and technical requirements for workers.

Center Rest and Follow Rest

Turning a slender shaft, when the length is more than 10 times the diameter, due to the insufficient rigidity of the workpiece itself, the workpiece will produce bending deformation under the influence of workpiece weight and cutting force, which will affect the processing accuracy. To use the additional auxiliary support - center frame or heel frame should be adopted at this time in order to prevent the bending of the workpiece.

The center rest is pressed on the bed, and the workpiece is supported by its three 120 degree supporting claws for processing. Because the center rest is compressed on the bed, so the big trailer can not exceed it, so when machining long workpieces, we need to machining one end firstly, then turn the head to install and then machining the other end.

Usually the central rest is used:

1. The piece has a large diameter and long length. It can not pass through the spindle hole of the machine tool.

2. Boring, boring or tapping at the end of the long rod.

3. Generally, it is used to machining the stepped shafts.

The follow rest is fixed on the lathe large trailer board for use. It moves with the cutter and adjusts the two supporting points to move just behind the cutter, so that the bearing points with cushions are supported on the surface just machined. The follow rest is used mainly for the machining of slender shafts which are not convenient to install on the head of the chuck.

When using the follow rest or the center rest, the supporting part of the workpiece should be the machined surface of the outer circle, and lubricated with oil. The rotational speed of the workpiece should not be too high, so that the friction between the workpiece and the supporting claw is too hot to burn or wear the supporting claw.

Center

Center is suitable for installing long axis workpieces for low speed finishing and semi-finishing. The center can be divided into front center and back center.

Front center: The center inserted in the spindle taper and rotated with the spindle. The front center rotates with the workpiece, and has no relative motion with the central hole and no friction.

Rear center: The center inserted into the sleeve of the tailstock of the lathe, and the rear apex is divided into fixed center and rotary center.

To clamp the workpiece with two centers, it is used mostly in the machining that needs to be clamped many times, and the same position is required. It should be placed some pieces, such as a soft shim, copper sheet etc. between the workpieces and the fastening screw in order to protect the machined surface.

Mandrel

The installation of mandrel is based on the inner hole of workpiece to ensure the position accuracy of parts. Small and medium-sized sleeves, pulleys and other parts can be installed by the mandrel. The manufacture of the mandrel is easy and the use is convenient, so it is widely used in the factory. Some disc and sleeve parts with high requirements of shape or coaxially need to be machined by the installation of the spindle. This should machining the holes firstly, then locate the holes and install them on the mandrel to machining the outer circle. According to the different requirements of shape, size and precision of the workpiece and the number of machined workpieces, different mandrel with different structures should be adopted.

◆ Small taper mandrel: When the length of the workpiece is larger than the workpiece aperture, a small taper mandrel with a slight taper can be used.

◆ Cylindrical mandrel: When the length of the workpiece is smaller than the aperture, it should be made into a cylindrical mandrel with nuts.

◆ Expansion mandrel: suitable for installation of small and medium-sized workpieces. By adjusting the conical screw, the workpiece is expanded and tightened with a little radial expansion at one end of the mandrel.

Disc Chuck

Disk chuck is often used to clamp workpieces with complex shapes. When fixing the workpiece on the disc chuck, it takes time to adjust the right position. At the same time, balancing iron and bending plate should be used to balance workpiece , so as to prevent vibration when rotating.

6.4 车刀

车刀

由于车削加工的内容不同,必须采用各种不同种类的车刀。车刀按其用途可分如下几类:

(1) 偏刀:用来车削工件的外圆、台阶和端面。
(2) 弯头车刀:用来车削工件的外圆、端面和倒角。
(3) 切断刀:用来切断工件或在工件上切出沟槽。
(4) 镗孔刀:用来镗削工件的内孔。
(5) 成形刀:用来车削工件台阶处的圆角和圆槽或车削特形面工件。
(6) 螺纹车刀:用来车削螺纹。

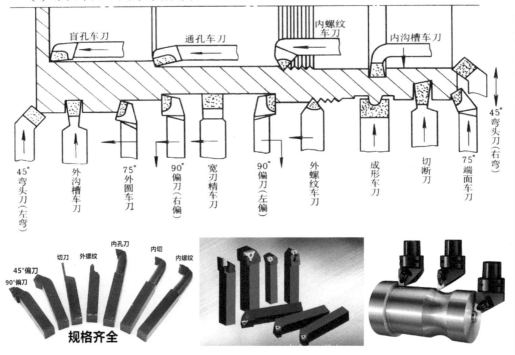

车刀的组成:

车刀由刀头(或刀片)和刀杆两部分组成。刀头是车刀的切削部分,刀杆是车刀的夹持部分。刀头部分由三面两刃一尖组成。

◆ 三面

(1) 前刀面(前面):切屑流出的表面。
(2) 主后刀面(后刀面、主后面):刀具上与工件的切削表面相对的表面。
(3) 副后刀面(副后面):刀具上与工件的已加工面相对的表面。

◆ 两刃

(1) 主切削刃(主刀刃):前刀面与主后刀面的交线,起主要的切削作用。

(2) 副切削刃 (副刀刃)：前刀面与副后刀面的交线，起辅助的切削作用。

◆ 刀尖

刀尖是主切削刃和副切削刃的交点。通常为一小圆弧。

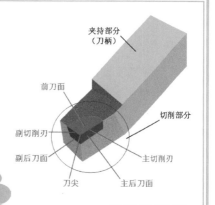

扫码了解车刀类型及车刀相关参数！

6.5 车削加工

车削加工的应用十分广泛。

车端面

车外圆

切槽

仿形车削

切槽

车螺纹

切槽

车端面槽

镗孔

是啊！让我们扫一扫二维码，通过视频来真实地感受一下车削加工吧！

第七章 铣削加工

7.1 概述

铣削加工是在铣床上使用旋转多刃刀具，对工件进行切削加工的方法。它是机械加工中最常用的加工方法之一。铣削是铣刀旋转作为主运动，铣刀或工件沿坐标方向的直线运动或回转运动是进给运动。

铣削加工的范围有哪些呢？

铣刀一般有几个齿同时参加切削，铣削能形成的工件型面有平面、槽、成形面、螺旋槽、齿轮和其他特殊型面。

看图 + 扫码！清楚又直观！

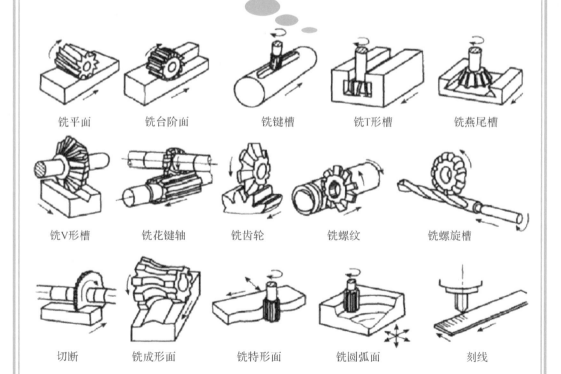

铣平面　　铣台阶面　　铣键槽　　铣T形槽　　铣燕尾槽

铣V形槽　　铣花键轴　　铣齿轮　　铣螺纹　　铣螺旋槽

切断　　铣成形面　　铣特形面　　铣圆弧面　　刻线

- 生产率较高。由于铣削是多齿刀具，铣削时有几个刀齿同时参加切削，总的切削宽度较大。铣削的主运动是铣刀的旋转，有利于告诉铣削，所以铣削的生产率一般比刨削高。

- 刀齿散热条件较好。铣刀刀齿在切离工件的一段时间内，可以得到一定的冷却，散热条件较好。但是，切入和切离时热和力的冲击将加速刀具的磨损，甚至可能引起硬质合金刀片的碎裂。

- 容易产生震动。由于铣削时参加切削的刀齿数以及在铣削时每个刀齿的切削厚度的变化，会引起切削力和切削面积的变化。因此，铣削过程不平稳，容易产生振动。铣削过程的不平稳性限制了铣削加工质量和生产率的进一步提高。

铣削加工工艺特点

Milling is the process of machining using rotary cutters to remove material by advancing a cutter into a workpiece. This may be done varying direction on one or several axes, cuter head speed, and pressure.Milling covers a wide variety of different operations and machines, on scales from small individual parts to large, heavy-duty gang milling operations. It is one of the most commonly used processes for machining custom parts to precise tolerances.

Milling can be done with a wide range of machine tools. The original class of machine tools for milling was the milling machine (often called a mill). After the advent of computer numerical control (CNC), milling machines evolved into machining centers: milling machines augmented by automatic tool changers, tool magazines or carousels, CNC capability, coolant systems, and enclosures. Milling centers are generally classified as vertical machining centers (VMCs) or horizontal machining centers (HMCs).

The integration of milling into turning environments, and vice versa, began with live tooling for lathes and the occasional use of mills for turning operations. This led to a new class of machine tools, multitasking machines (MTMs), which are purpose-built to facilitate milling and turning within the same work envelope.

7.2 铣床

铣床按布局形式和适用范围分类

下面我们举例介绍一下!

卧式铣床

X6132型卧式万能铣床

为什么要叫卧式铣床呢?难道机床躺下啦?

当然不是啦!因为铣床的主轴轴线与工作台面平行,主轴呈横卧位置,所以称作卧式铣床。让我们一起看看它的结构吧!

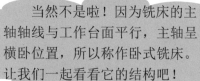
卧式铣床是如何完成切削的?它的加工范围有哪些?

铣削时将铣刀安装在与主轴相连接的刀杆上,随主轴作旋转运动,被切削工件装夹在工作台面上对铣刀作相对进给运动,从而完成切削工作。

机械加工、成型和安装工艺

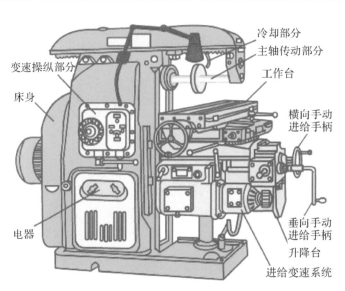

X62卧式铣床的结构

卧式铣床加工范围很大，可以加工沟槽、平面、成形面、螺旋槽等。根据加工范围的大小，卧式铣床又可分为一般卧式铣床（平铣）和卧式万能铣床。

卧式万能铣床的结构与一般卧式铣床不同，其纵向工作台与横向工作台之间有一转盘，并具有回转刻度线。使用时，可以按照需要在±45°范围内扳转角度，适用于圆盘铣刀加工螺旋槽等工件。同时，卧式万能铣床还带有较多附件，因而于加工范围较广。由于卧式万能铣床具有以上优点，所以得到广泛的应用。

立式铣床

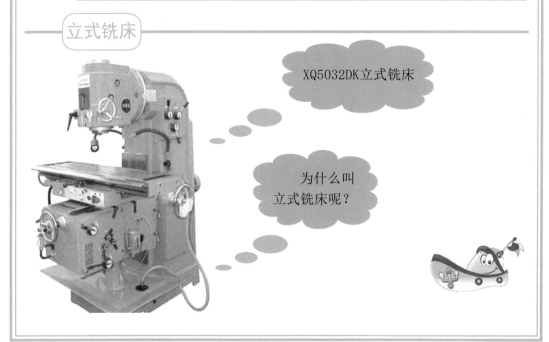

XQ5032DK立式铣床

为什么叫立式铣床呢？

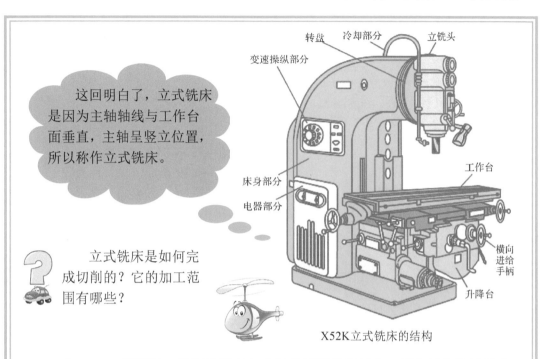

这回明白了，立式铣床是因为主轴轴线与工作台面垂直，主轴呈竖立位置，所以称作立式铣床。

 立式铣床是如何完成切削的？它的加工范围有哪些？

X52K立式铣床的结构

铣削时，铣刀安装在与主轴相连接的刀轴上，绕主轴作旋转运动，被切削工件装夹在工作台上，对铣刀作相对运动，完成切削过程。

立式铣床加工范围大，通常可以应用面铣刀、立铣刀、成形铣刀等铣削各种沟槽、表面；另外，利用机床附件，如回转工作台、分度头，还可以加工圆弧、曲线外形、齿轮、螺旋槽、离合器等较复杂的零件；当生产批量较大时，在立铣上采用硬质合金刀具进行高速铣削，可以大大提高生产率。

立式铣床与卧式铣床相比，在操作上还具有观察清楚、检查调整方便等特点。

立式铣床按其立铣头的不同结构，又可分为两种：

(1) 立铣头与机床床身是一个整体，因此刚性较好，但加工范围较小。

(2) 立铣头与机床床身之间有一回转盘，盘上有刻度线，主轴随立铣头可扳转一定角度，以适应铣削各种角度面、椭圆孔等工件。由于立铣头可以回转，所以生产中应用较广。

万能工具铣床

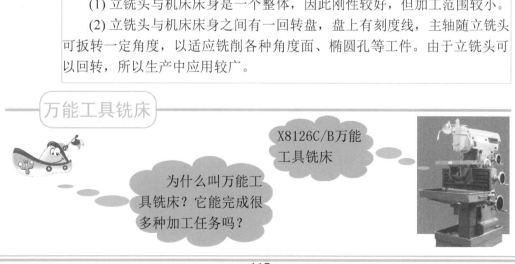

X8126C/B万能工具铣床

为什么叫万能工具铣床？它能完成很多种加工任务吗？

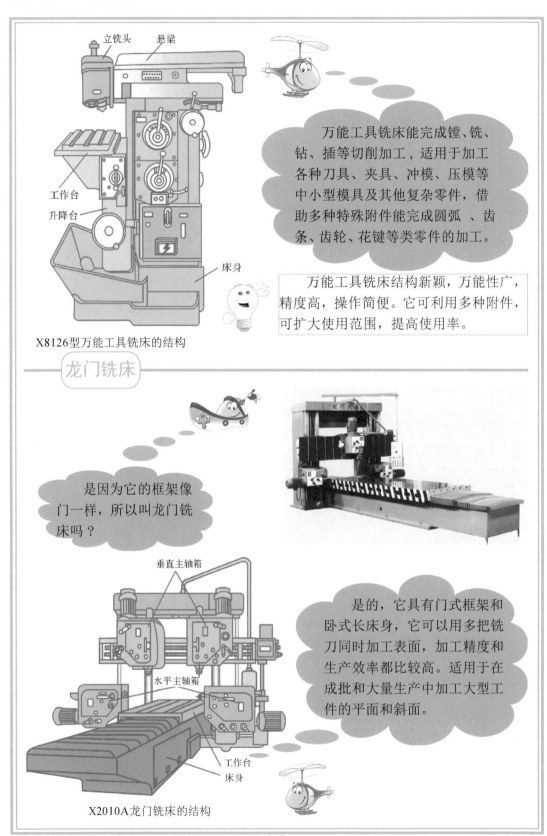

按形式划分

- 按龙门架是否移动，分为龙门固定工作台移动式（约占90%以上）和龙门移动式（又称桥式）。桥式龙门铣的特点是占地面积小，承载能力大。龙门架行程可达20米，便于加工特长或特重的工件。
- 按横梁是否在立柱上运动，分为动梁式和定梁式两种；横梁在高架床身上移动称为高架式。

按系统划分

- 分为普通型和数控型，还有厂家开发出普通铣和数控铣一键式转换的数字智能化龙门铣，更为灵活实用。

按能力划分

- 分为轻型、中型和重型（超重型）龙门铣。

龙门铣床的种类

龙门铣床的特点

当然啦！满足你的要求！

扫码是不是会了解得更清楚呢？

- 机床刚性高
- 机床生产效率高
- 机床操作方便
- 机床结构简单
- 机床整体性能全面

 介绍了这么多种铣床，那铣床的主要部件都有哪些呢？它们的作用又是什么呢？

铣床一般由以下几部分组成：床身、横梁及挂架、主轴、升降台、横向工作台、纵向工作台、转台、主轴变速机构、进给变速机构等。

◆床身：机床的主体，用来安装和支承机床其他部件，其刚性、强度和精度对铣削效率和加工质量影响很大。因此，在设计时，都将床身内壁加肋条，以增加刚性和强度。

机械加工、成型和安装工艺

◆**横梁及挂架**：横梁安装在床身顶部，挂架安装在横梁上，主要是支持刀轴外端，增加刀轴的刚性。对于卧式铣床来说，其安装在导轨上，可根据工作要求沿导轨作前后移动。

◆**主轴**：用于安装或通过刀杆安装铣刀，并带动铣刀旋转，其前端是带锥孔的空心轴，其锥度一般是 7:24，用来安装铣刀或铣刀杆。

◆**升降台**：它是工作台的支座，在升降台上安装着铣床的纵向工作台、横向工作台和转台。升降台安装在床身前侧的垂直导轨上，中部有丝杠与底座螺母相连接，主要作用是支撑工作台并带动其作上、下移动，即铣削时的垂直进给运动。进给电动机变速、操纵机构等都安装在升降台上。升降台的刚性和精度要求都很高。

◆**横向工作台**：沿升降台水平导轨作横向进给运动。位于纵向工作台下方，以带动纵向工作台作前后移动。

◆**纵向工作台**：用来安装工件或夹具，并带着工件作纵向进给运动。其上有三条 T 形槽，用来安装压板螺栓(T 形螺栓)。这三条 T 形槽中，当中一条精度较高，其余两条精度较低。工作台前侧面有一条小 T 形槽，用来安装行程挡铁。纵向工作台台面的宽度是标志铣床大小的主要规格。

◆**转台**：万能铣床在纵向工作台和横向工作台之间，还有一层转台，其唯一作用是能将纵向工作台在水平面内回转一个正、反不超过 45 度的角度，以便铣削螺旋槽。有无转台是区分万能卧式铣床和一般卧式铣床的标志。

◆**主轴变速机构**：它安装在床身内，作用是将主电动机的额定转速通过齿轮变速，变换成不同转速，传递给主轴，以适应铣削的需要。

◆**进给变速机构**：安装在升降台内，作用是将进给电动机的额定转速通过齿轮变速传递给进给机构，实现工作台移动的不同速度以适应铣削要求。

扫一扫了解铣床结构、执行机构及其运动方式。

Horizontal milling machine

How does a horizontal milling machine complete cutting? What is its processing range?

In milling, the milling cutter is installed on the cutter bar connected with the spindle, rotates with the spindle, and is clamped by the cutting workpiece on the worktable to make relative feed motion to the milling cutter, thus completing the cutting work. Horizontal milling machine can process grooves, planes, forming surfaces, spiral grooves and so on. According to the size of processing range, horizontal milling machine can be divided into general horizontal milling machine (flat milling) and horizontal universal milling machine.

The structure of horizontal universal milling machine is different from that of general horizontal milling machine. There is a turntable between the vertical table and the horizontal table, and there is a rotary calibration line. When in use, the angle can be turned in the range of +45 degrees as required, and the disc milling cutter is suitable for processing spiral grooves and other workpieces. At the same time, the horizontal universal milling machine also has more accessories, so the processing range is wider. Horizontal universal milling machine has the above advantages, so it has been widely used.

Vertical milling machine

How does a vertical milling machine complete cutting? What is its processing range?

In milling, the milling cutter is installed on the tool axis connected with the spindle, rotates around the spindle, is clamped on the worktable by the cutting workpiece, works relative motion on the milling cutter, and completes the cutting process.

Vertical milling machine has a wide range of processing. It can usually use face milling cutter, end milling cutter, forming milling cutter and so on to milling various grooves and surfaces. In addition, it can also process more complex parts such as arc, curve shape, gear, spiral groove and clutch by using machine tool accessories, such as rotary table and indexing head. In large batches, high speed milling with carbide cutters in end milling can greatly improve productivity.

Compared with horizontal milling machine, vertical milling machine has the characteristics of clear observation, convenient inspection and adjustment.

Vertical milling machine can be divided into two kinds according to the different structures of its end milling head:

(1) End milling head and machine tool bed are a whole, so the rigidity is better, but the processing range is smaller.

(2) There is a turntable between the end milling head and the machine tool bed. The dial has a scale line. The spindle can be rotated at a certain angle with the end milling head in line with milling various angles, elliptical holes and other workpieces. Because the end milling head can be rotated, it is widely used in production.

Vertical mill

In the **vertical mill** the spindle axis is vertically oriented. Milling cutters are held in the spindle and rotate on its axis. The spindle can generally be extended (or the table can be raised/lowered, giving the same effect), allowing plunge cuts and drilling. There are two subcategories of vertical mills: the bed mill and the turret mill.

◆ A **turret mill** has a stationary spindle and the table is moved both perpendicular and parallel to the spindle axis to accomplish cutting. The most common example of this type is the Bridgeport, described below. Turret mills often have a quill which allows the milling cutter to be raised and lowered in a manner similar to a drill press. This type of machine provides two methods of cutting in the vertical (Z) direction: by raising or lowering the quill, and by moving the knee.

◆ In the **bed mill**, however, the table moves only perpendicular to the spindle's axis, while the spindle itself moves parallel to its own axis.

Turret mills are generally considered by some to be more versatile of the two designs. However, turret mills are only practical as long as the machine remains relatively small. As machine size increases, moving the knee up and down requires considerable effort and it also becomes difficult to reach the quill feed handle (if equipped). Therefore, larger milling machines are usually of the bed type.

A third type also exists, a lighter machine, called a mill-drill, which is a close relative of the vertical mill and quite popular with hobbyists. A mill-drill is similar in basic configuration to a small drill press, but equipped with an X-Y table. They also typically use more powerful motors than a comparably sized drill press, with potentiometer-controlled speed and generally have more heavy-duty spindle bearings than a drill press to deal with the lateral loading on the spindle that is created by a milling operation. A mill drill also typically raises and lowers the entire head, including motor, often on a dovetailed vertical, where a drill press motor remains stationary, while the arbor raises and lowers within a driving collar. Other differences that separate a mill-drill from a drill press may be a fine tuning adjustment for the Z-axis, a more precise depth stop, the capability to lock the X, Y or Z axis, and often a system of tilting the head or the entire vertical column and powerhead assembly to allow angled cutting. Aside from size and precision, the principal difference between these hobby-type machines and larger true vertical mills is that the X-Y table is at a fixed elevation; the Z-axis is controlled in basically the same fashion as drill press, where a larger vertical or knee mill has a vertically fixed milling head, and changes the X-Y table elevation. As well, a mill-drill often uses a standard drill press-type Jacob's chuck, rather than an internally tapered arbor that accepts collets. These are frequently of lower quality than other types of machines, but still fill the hobby role well because they tend to be benchtop machines with small footprints and modest price tags.

Horizontal milling machine

A horizontal mill has the same sort but the cutters are mounted on a horizontal spindle (see Arbor milling) across the table. Many horizontal mills also feature a built-in rotary table that allows milling at various angles; this feature is called a *universal table.* While endmills and the other types of tools available to a vertical mill may be used in a horizontal mill, their real advantage lies in arbor-mounted cutters, called side and face mills, which have a cross section rather like a circular saw, but are generally wider and smaller in diameter. Because the cutters have good support from the arbor and have a larger cross-sectional area than an end mill, quite heavy cuts can be taken enabling rapid material removal rates. These are used to mill grooves and slots. Plain mills are used to shape flat surfaces. Several cutters may be ganged together on the arbor to mill a complex shape of slots and planes. Special cutters can also cut grooves, bevels, radii, or indeed any section desired. These specialty cutters tend to be expensive. Simplex mills have one spindle, and duplex mills have two. It is also easier to cut gears on a horizontal mill. Some horizontal milling machines are equipped with a power-take-off provision on the table. This allows the table feed to be synchronized to a rotary fixture, enabling the milling of spiral features such as hypoid gears.

Universal tool milling machine

Universal tool milling machine can complete boring, milling, drilling, insertion and other cutting, suitable for machining various cutters, fixtures, dies, pressing dies and other small and medium-sized molds and other complex parts, with the help of a variety of special accessories which can complete arc, rack, gear, spline and other parts processing.

Universal tool milling machine has novel structure, wide versatility, high precision and simple operation. It can use a variety of accessories, expand the scope of use, and improve the utilization rate.

gantry milling machine

It has a door frame and a long horizontal bed. It can use multiple milling cutters to machining the surface at the same time. The machining accuracy and production efficiency are relatively high. It is suitable for machining the plane and incline of large-scale work in batch and master production.

What are the main parts of the milling machine?

Milling machine generally consists of the following parts: bed, cross beam and hanger, spindle, lifting platform, horizontal table, vertical table, turntable, spindle transmission mechanism, feed transmission mechanism.

◆ Bed: The main body of a machine tool is used to install and support other parts of the machine tool. Its rigidity, strength and precision have a great impact on milling efficiency and processing quality. Therefore, the inner wall is added with the aid strip to increase the rigidity and strength in the design.

◆ Beams and hangers: The cross beams are installed on the top of the bed, and the hangers are installed on the cross beams, which mainly support the outer end of the cutter shaft and increase the rigidity of the cutter shaft. For the horizontal milling machine, it is installed on the rail and can move forward and backward along the guide according to the working requirements.

◆ Spindle: Used to install or install milling cutter through cutter rod, and drive the milling cutter to rotate. Its front end is a hollow shaft with taper hole. Its taper is usually 7:24, which is used to install milling cutter or milling cutter rod.

◆ Lifting table: It is the support of the worktable. The vertical worktable, the horizontal worktable and the turntable of the milling machine are installed on the lifting table. The lifting platform is installed on the vertical guide rail at the front of the bed, and the screw is connected with the base nut in the middle. It mainly supports the worktable and drives it to move up and down, that is, the vertical feed movement in milling. The variable speed and control mechanism of the feed motor are all installed on the lifting platform. Rigidity and accuracy requirements of lifting platform are very high.

◆ Horizontal Workbench: Horizontal feed motion along the horizontal guide rail of the lifting platform. Located under the vertical worktable to drive the vertical worktable forward and backward.

◆ Longitudinal worktable: It is used to install workpiece or fixture, and take workpiece for longitudinal feed movement. There are three T-grooves on it, which are used to install pressure plate bolts (T-bolts). Among the three T-grooves, one has higher accuracy and the other two have lower accuracy. There is a small T-shaped groove on the front side of the worktable to install the travel iron shield. The width of the vertical table is the main specification to mark the size of the milling machine.

◆ Turntable: The universal milling machine has a layer of turntable between the vertical and horizontal worktables. Its only function is to turn the vertical worktable in the horizontal plane at an angle of no more than 45 degrees in order to milling spiral grooves. Whether there is a turntable or not is the only sign to distinguish universal horizontal milling machine from general horizontal milling machine.

◆ Spindle speed change mechanism: It is installed in the bed, the role is to change the rated speed of the main motor through gear speed change, into different speed, transmission to the spindle, to meet the needs of milling.

◆ Feed Variable Speed Mechanism: Installed in the lifting platform, the function is to transfer the rated speed of the feed motor to the feed mechanism through the gear variable speed, so as to realize the different speed of the table movement to meet the requirements of milling.

7.3 铣床附件

铣床附件种类很多，我们主要介绍四种附件，下面让我们来学习一下关于铣床附件的知识吧！

万能分度头

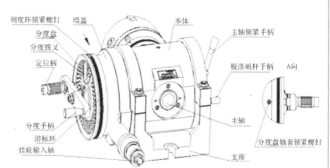

万能分度头的结构及使用方法

利用分度盘、分度拨叉、分度手柄、游标环以及定位销，将装卡在顶尖间或卡盘上的工件分成任意角度，可将圆周任意等分，辅助机床利用不同形状的刀具进行各种沟槽、正齿轮、螺旋正齿轮、阿基米德螺线凸轮等的加工工作。

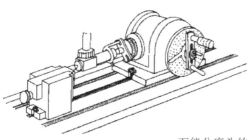

万能分度头的使用方法

万能分度头的作用

- 能将工件作任意周圆等分或作直线移距分度。
- 可将工件装夹成所需要的角度(垂直、水平或倾斜)。
- 通过交换齿轮，可使分度头与工作台传动系统连接，使分度头主轴随工作台的进给运动作连续转动，以铣削螺旋面和凸轮面。

回转工作台

回转工作台是指带有可转动的台面,用以装夹工件并实现回转和分度定位的机床附件,简称转台或第四轴。

用回转工作台装卡工件,可以扩大工艺性,缩短辅助时间。用它可作镗孔,直线及环形平面的铣削或磨削等工序的调整、进给、分度、换向等运动。备有较宽的调速范围,可获得较佳进给量。

万能铣头

万能铣头也叫万向铣头,是指机床刀具主轴可在水平和垂直两个平面内回转的铣头;从机床坐标系来看,就是机床刀具输出轴能够围绕机床 Z 轴和 X 轴(或 Y 轴)旋转的铣头,其中围绕机床 Z 轴的轴叫 C 轴,围绕机床 X 轴的轴叫 A 轴,从而使机床具备五个坐标轴。

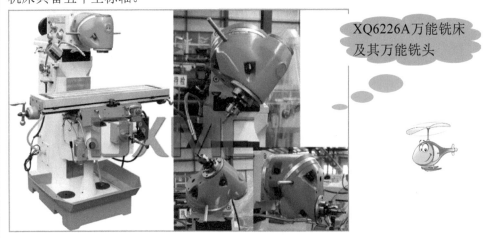

XQ6226A万能铣床及其万能铣头

第三篇　切削加工与特种加工

万能铣头是卧式升降台铣床的主要附件，用以扩大铣床的使用范围和功能。万能铣头主轴可以在相互垂直的两个回转面内回转，不仅能完成立铣、平铣工作，而且可以在工件一次装卡中，进行各种角度的多面、多棱、多槽的铣削。

平口钳

平口钳又名机用虎钳，是一种通用夹具，常用于安装小型工件。它是铣床、钻床的随机附件。将其固定在机床工作台上，用来夹持工件进行切削加工。一般用于小型较规则的零件，如较方正的板块类零件、盘套类零件、轴类零件和小型支架等。

milling machine accessory

universal indexing head

Using indexing rings and verniers, positioning pins, indexing plates and exchange gears, the workpiece clamped between the apex or chuck can be divided into arbitrary angles, and the circumference can be divided into arbitrary equal parts. The auxiliary machine tool uses various shapes of cutters to carry out various grooves, spur gears, Archimedes spiral cams and so on.

rotating table

Rotating table refers to machine tool accessories with rotatable table to clamp workpieces and achieve rotary and indexing positioning, known as turntable or the fourth axis.

By using it to fix the workpiece, the workability can be enlarged and the auxiliary time can be shortened. It can be used for adjusting, feeding, indexing and commutation of boring, milling or grinding of straight line and annular plane. With a wide range of speed regulation, a better feed rate can be obtained.

universal milling head

Universal milling head, also known as universal milling head, refers to the milling head whose tool axes can rotate in horizontal and vertical planes; from the machine tool coordinate system, it is the milling head whose tool axes can rotate around Z axis and X axis (or Y axis) of the machine tool, in which the axes around Z axis of the machine tool are called C axis, and the axes around X axis of the machine tool are called A axis. The machine tool has five coordinate axes.

Universal milling head is the main accessory of horizontal lifting table milling machine, which is used to expand the scope and function of milling machine. The spindle of universal milling head can rotate in two revolving planes which are perpendicular to each other. It can not only finish the work of end milling and flat milling, but also carry out the multi-sided, multi-edged and multi-groove milling of various angles in one loading of workpiece.

bench vise

Bench vise, also known as machine vise, is a kind of universal fixture, often used to install small workpieces. It is a random accessory of milling machine and drilling machine. It is fixed on the machine tool worktable to clamp the workpiece for cutting, Generally used for small and regular parts, such as rectangular plate parts, disc sleeve parts, shaft parts and small brackets.

7.4 铣刀

铣 刀

铣刀是用于铣削加工的、具有一个或多个刀齿的旋转刀具。工作时各刀齿依次间歇地切去工件的余量。铣刀主要用于在铣床上加工平面、台阶、沟槽、成形表面和切断工件等。

扫码看铣刀分类及铣刀角度介绍

机械加工、成型和安装工艺

铣刀类型	用途
圆柱形铣刀	·用于卧式铣床上加工平面。
面铣刀	·用于立式铣床、端面铣床或龙门铣床上加工平面。
立铣刀	·主要用于平面、凹槽、台阶面和仿形铣削。
三面刃铣刀	·用于加工各种沟槽和台阶面。
角度铣刀	·用于铣成型角度的平面，或加工相应角度的槽。
锯片铣刀	·用于加工深槽和切断工件。
T形铣刀	·用来铣T形槽。

各类铣刀的用途

Milling cutter

It is a rotary tool with one or more cutter teeth for milling. During the work, each cutter teeth cut off the workpiece allowance intermittently in turn. Milling cutter is mainly used to process plane, step, groove, forming surface and cutting workpiece on milling machine.

Milling cutters are cutting tools typically used in milling machines or machining centres to perform milling operations (and occasionally in other machine tools). They remove material by their movement within the machine (e.g., a ball nose mill) or directly from the cutter's shape (e.g., a form tool such as a hobbing cutter).

Types

1. End Mill

End mills (middle row in image) are those tools which have cutting teeth at one end, as well as on the sides. The words end *mill* are generally used to refer to flat bottomed cutters, but also include rounded cutters (referred to as *ball nosed*) and radiused cutters (referred to as *bull nose*, or *torus*). They are usually made from high speed steel or cemented carbide, and have one or more flutes. They are the most common tool used in a vertical mill.

2. Roughing end mill

Roughing end mills quickly remove large amounts of material. This kind of end mill utilizes a wavy tooth form cut on the periphery. These wavy teeth form many successive cutting edges producing many small chips, resulting in a relatively rough surface finish, but the swarf takes the form of short thin sections and is more manageable than a thicker more riboon-like section. During cutting, multiple teeth are in simultaneous contact with the workpiece reducing chatter and vibration. Rapid stock removal with heavy milling cuts is sometimes called *hogging*. Roughing end mills are also sometimes known as "rippa" or "ripper" cutters.

3.Ball cutter

Ball nose cutters or ball end mills (lower row in image) are similar to slot drills, but the end of the cutters are hemispherical. They are ideal for machining 3-dimensional contoured shapes in machining centres, for example in moulds and dies. They are sometimes called *ball mills* in shop-floor slang, despite the fact that that term also has another meaning. They are also used to add a radius between perpendicular faces to reduce stress concentrations.

There is also a term *bull nose* cutter, which refers to a cutter having a corner radius that is fairly large, although less than the spherical radius (half the cutter diameter) of a ball mill; for example, a 20-mm diameter cutter with a 2-mm radius corner. This usage is analogous to the term *bull nose center* referring to lathe centers with truncated cones;

in both cases, the silhouette is essentially a rectangle with its corners truncated (by either a chamfer or radius Don).

4. Slab mill

Slab mills are used either by themselves or in gang milling operations on manual horizontal or universal milling machines to machine large broad surfaces quickly. They have been superseded by the use of cemented carbide-tipped face mills which are then used in vertical mills or machining centres.

5. Side-and-face cutter

The side-and-face cutter is designed with cutting teeth on its side as well as its circumference. They are made in varying diameters and widths depending on the application. The teeth on the side allow the cutter to make *unbalanced cuts* (cutting on one side only) without deflecting the cutter as would happen with a slitting saw or slot cutter (no side teeth).

Cutters of this form factor were the earliest milling cutters developed. From the 1810s to at least the 1880s they were the most common form of milling cutter, whereas today that distinction probably goes to end mills.

6. Involute gear cutter

There are 8 cutters (excluding the rare half sizes) that will cut gears from 12 teeth through to a rack (infinite diameter).

7. Hob

These cutters are a type of form tool and are used in hobbing machines to generate gears. A cross section of the cutter's tooth will generate the required shape on the workpiece, once set to the appropriate conditions (blank size). A hobbing machine is a specialised milling machine.

8. Thread mill

Whereas a hob engages the work much as a mating gear would (and cuts the blank progressively until it reaches final shape), a thread milling cutter operates much like an endmill, traveling around the work in a helical interpolation.

9. Face mill

A face mill is a cutter designed for facing as opposed to e.g., creating a pocket (end mills). The cutting edges of face mills are always located along its sides. As such it must always cut in a horizontal direction at a given depth coming from outside the stock. Multiple teeth distribute the chip load, and since the teeth are normally disposable carbide inserts, this combination allows for very large and efficient face milling.

10. Fly cutter

A fly cutter is composed of a body into which one or two tool bits are inserted. As the entire unit rotates, the tool bits take broad, shallow facing cuts. The purpose that fly cutters are analogous to face mills is that face milling and their individual cutters are replaceable. Face mills are more ideal in various respects (e.g., rigidity, indexability of inserts without disturbing effective cutter diameter or tool length offset, depth-of-cut capability), but tend to be expensive, whereas fly cutters are very inexpensive.

Most fly cutters simply have a cylindrical center body that holds one tool bit. It is usually a standard left-hand turning tool that is held at an angle of 30 to 60 degrees. Fly cutters with two tool bits have no "official" name but are often called double fly cutters, double-end fly cutters, or fly bars. The latter name reflects that they often take the form of a bar of steel with a tool bit fastened on each end. Often these bits will be mounted at right angles to the bar's main axis, and the cutting geometry is supplied by using a standard right-hand turning tool.

Regular fly cutters (one tool bit, swept diameter usually less than 100 mm) are widely sold in machinists' tooling catalogs. Fly bars are rarely sold commercially; they are usually made by the user. Fly bars are perhaps a bit more dangerous to use than endmills and regular fly cutters because of their larger swing. As one machinist put it, running a fly bar is like "running a lawn mower without the deck", that is, the exposed swinging cutter is a rather large opportunity to take in nearby hand tools, rags, fingers, and so on. However, given that a machinist can never be careless with impunity around rotating cutters or workpieces, this just means using the same care as always except with slightly higher stakes. Well-made fly bars in conscientious hands give years of trouble-free, cost-effective service for the facing off of large polygonal workpieces such as die/mold blocks.

11. Woodru cutter

Woodruff cutters are used to cut the keyway for a woodruff key.

12. Hollow mill

Hollow milling cutters, more often called simply *hollow mills*, are essentially "inside-out endmills". They are shaped like a piece of pipe (but with thicker walls), with their cutting edges on the inside surface. They are used on turret lathes and screw machines as an alternative to turning with a box tool, or on milling machines or drill presses to finish a cylindrical boss (such as a trunnion).

13. Dovetail cutter

A dovetail cutter is an end mill whose form leaves behind a dovetail slot, such as often forms the ways of a machine tool.

14. Shell mill

(1) Modular principle

A shell mill is any of various milling cutters (typically a face mill or endmill) whose construction takes a modular form, with the shank (arbor) made separately from the body of the cutter, which is called a "shell" and attaches to the shank/arbor via any of several standardized joining methods.

This modular style of construction is appropriate for large milling cutters for about the same reason that large diesel engines use separate pieces for each cylinder and head whereas a smaller engine would use one integrated casting. Two reasons are that (1) for the maker it is more practical (and thus less expensive) to make the individual pieces as separate endeavors than to machine all their features in relation to each other while the whole unit is integrated (which would require a larger machine tool work envelope); and (2) the user can change some pieces while keeping other pieces stay the same (rather than changing the whole unit). One arbor (at a hypothetical price of USD100) can serve for various shells at different times. Thus 5 different milling cutters may require only USD100 worth of arbor cost, rather than USD500, as long as the workflow of the shop does not require them all to be set up simultaneously. It is also possible that a crashed tool scraps only the shell rather than both the shell and arbor. This would be like crashing a "regular" endmill and being able to reuse the shank rather than losing it along with the flutes.

Most shell mills made today use indexable inserts for the cutting edges—thus shank, body, and cutting edges are all modular components.

(2) Mounting methods

There are several common standardized methods of mounting shell mills to their arbors. They overlap somewhat (not entirely) with the analogous joining of lathe chucks to the spindle nose.

The most common type of joint between shell and arbor involves a fairly large cylindrical feature at center (to locate the shell concentric to the arbor) and two driving lugs or tangs that drive the shell with a positive engagement (like a dog clutch). Within the central cylindrical area, one or several socket head cap screws fasten the shell to the arbor.

Another type of shell fastening is simply a large-diameter fine thread. The shell then screws onto the arbor just as old-style lathe chuck backplates screw onto the lathe's spindle nose. This method is commonly used on the 2" or 3" boring heads used on knee mills. As the threaded-spindle-nose lathe chucks, this style of mounting requires that the cutter only take cuts in one rotary direction. Usually (i.e., with right-hand helix orientation) this means only M03, never M04, or in pre-CNC terminology, "only forward, never reverse". One could use a left-hand thread if one needed a mode of use involving the opposite directions (i.e., only M04, never M03).

7.5 铣削要素及铣削方式

铣削用量

铣削用量包括切削速度、进给量、背吃刀量（切削深度）和侧吃刀量（铣削宽度）四个要素。

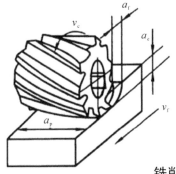 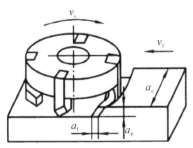

铣削要素表

铣削要素		含 义	符号	单位
切削速度		铣刀主运动的瞬时线速度	v_c	m/min
进给量	每齿进给量	铣削时工件与铣刀在进给方向上的相对位移量，有三种表示方法：每齿进给量、每转进给量、进给速度	铣刀每转一个刀齿时，工件与铣刀沿进给方向的相对位移量	f_z	mm/z
	每转进给量		铣刀每转一转时，工件与铣刀沿进给方向的相对位移量	f	mm/r
	进给速度		单位时间内工件与铣刀沿进给方向的相对位移量	v_f	mm/min
背吃刀量(切削深度)		平行于铣刀轴线方向测量的切削层尺寸	a_p	mm
侧吃刀量（切削宽度）		垂直于铣刀轴线方向测量的切削层尺寸	a_e	mm

注：1. 切削速度 $v_c = \dfrac{\pi d n}{1000}$

　　d——铣刀直径（mm）；n——铣刀每分钟转数（r/min）

2. 进给量中的每齿进给量、每转进给量、进给速度之间的关系为

　　$v_f = nf = nzf_z$

　　z——铣刀齿数；n——铣刀每分钟转速（r/min）

铣削方式

铣削方式是指铣削时铣刀相对于工件的运动和位置关系。它对铣刀寿命、工件表面粗糙度、铣削过程平稳性及生产率都有较大的影响。

铣削方式分为圆周铣削和端面铣削

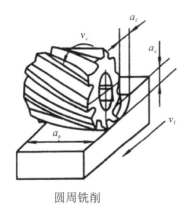

圆周铣削

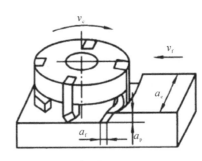

端面铣削

圆周铣削和端面铣削介绍

圆周铣削（周铣）	端面铣削（端铣）
用铣刀圆周上的切削刃来铣削工件，铣刀的回转轴线与被加工表面平行	用铣刀端面上的切削刃来铣削工件，铣刀的回转轴线与被加工表面垂直
通常只在卧式铣床上进行	端铣一般在立式铣床上进行，也可在其他形式的铣床上进行
只有主刃参加切削，无副刃，所以加工后的表面粗糙度较大	主刃副刃同时参加切削，且副刃有修光作用，所以加工后的表面粗糙度较小
周铣时主轴刚性差，生产率较低，适于在中小批生产中铣削狭长的平面、键槽及某些成形表面和组合表面	端铣时主轴刚性好，并且面铣刀易于采用硬质合金可转位刀片，因而采用切削用量大，生产率较高，适于在大批大量生产中铣削宽大平面

圆周铣削和端面铣削比较

比 较 内 容	周铣	端铣
有无修光刃 / 工件表面质量	无/差	有/好
刀杆刚度 / 切削振动	小/大	大/小
同时参加切削的刀齿 / 切削平稳性	少/差	多/好
易否镶嵌硬质合金刀片 / 刀具耐用度	难/低	易/高
生产率 / 加工范围	低/广	高/较小

周铣的铣削方式又分为：顺铣和逆铣。

周铣和端铣的比较

用圆柱铣刀铣削时,其铣削方式可分为顺铣和逆铣两种。

当工件的进给方向与圆柱铣刀刀尖圆和已加工平面的切点处的切削速度的方向相反时称为逆铣,反之为顺铣。

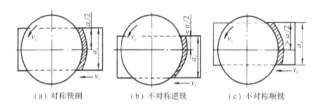

(a) 对称铣削　　(b) 不对称逆铣　　(c) 不对称顺铣

比较内容	顺铣	逆铣
工件夹紧程度/切削过程稳定性	好	差
刀具磨损	小	大
工作台丝杠和螺母有无间隙	有	无
由工作台窜动引起的质量事故	多	少
加工对象	精加工	粗加工

顺铣和逆铣的比较

根据铣刀与工件相对位置的不同,端铣可以分为对称铣与不对称铣两种方式。

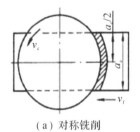  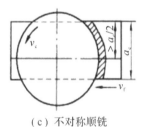

(a) 对称铣削　　(b) 不对称逆铣　　(c) 不对称顺铣

对称铣削与不对称铣削介绍

	对称铣削	铣削时铣刀轴线与工件铣削宽度对称中心线重合
不对称铣削	不对称逆铣	铣削时铣刀轴线与工件铣削宽度对称中心线不重合,若切入时的切削厚度小于切出时的切削厚度,称为不对称逆铣
	不对称顺铣	铣削时铣刀轴线与工件铣削宽度对称中心线不重合,若切入时的切削厚度大于切出时的切削厚度,称为不对称顺铣

铣削加工实例,就在扫一扫!

Using a milling cutter
Chip formation

Although there are many different types of milling cutter, understanding chip formation is fundamental to the use of any of them. As the milling cutter rotates, the material to be cut is fed into it, and each tooth of the cutter cuts away a small chip of material. Achieving the correct size of chip is of critical importance. The size of this chip depends on several variables.

Surface cutting speed (Vc)

This is the speed at which each tooth cuts through the material as the tool spins. This is measured either in metres per minute in metric countries, or surface feet per minute (SFM) in America. Typical values for cutting speed are 10 m/min to 60 m/min for some steels, and 100 m/min and 600 m/min for aluminum. This should not be confused with the feed rate. This value is also known as "tangential velocity."

Spindle speed (S)

This is the rotation speed of the tool, and is measured in revolutions per minute (rpm). Typical values are from hundreds of rpm, up to tens of thousands of rpm.

Diameter of the tool (D)
Number of teeth (z)
Feed per tooth (Fz)

This is the distance the material is fed into the cutter as each tooth rotates. This value is the size of the deepest cut the tooth will make.

Feed rate (F)

This is the speed at which the material is fed into the cutter. Typical values range from 20 mm/min to 5000 mm/min.

Depth of cut

This is how deep the tool is under the surface of the material being cut (not shown on the diagram). This will be the height of the chip produced. Typically, the depth of cut will be less than or equal to the diameter of the cutting tool.

第八章 磨削加工

8.1 概述

磨削加工的发展

磨削是人类自古以来就使用的一种古老技术，在旧石器时代磨制石器就是在使用这种技术。随后，金属器具的使用促进了研磨技术的发展。

18世纪30年代，为了适应钟表、自行车、缝纫机和枪械等零件的加工，英国、德国和美国分别研制出使用天然磨料砂轮的磨床。这些磨床是在当时现成的机床如车床、刨床等上面加装磨头改制而成的，它们结构简单，刚度低，磨削时易产生振动，要求操作工人要有很高的技艺才能磨出精密的工件。

1876年，万能外圆磨床在巴黎博览会展出，这是首次具有现代磨床基本特征的机械展现在人们面前。

1883年，美国布朗－夏普公司制成磨头装在立柱上，工作台作往复移动的平面磨床。

1900年前后，人造磨料的发展和液压传动的应用，对磨床的发展有很大的推动作用。随着近代工业特别是汽车工业的发展，各种不同类型的磨床相继问世。例如20世纪初，先后研制出加工气缸体的行星内圆磨床、曲轴磨床、凸轮轴磨床和带电磁吸盘的活塞环磨床等。

自动测量装置于1908年开始应用到磨床上，到了1920年前后，无心磨床、双端面磨床、轧辊磨床、导轨磨床、珩磨机和超精加工机床等相继制成使用；20世纪50年代又出现了可作镜面磨削的高精度外圆磨床；60年代末又出现了砂轮线速度达60～80米/秒的高速磨床和大切深、缓进给磨削平面磨床；70年代，采用微处理机的数字控制和适应控制等技术在磨床上得到了广泛的应用。

机械加工、成型和安装工艺

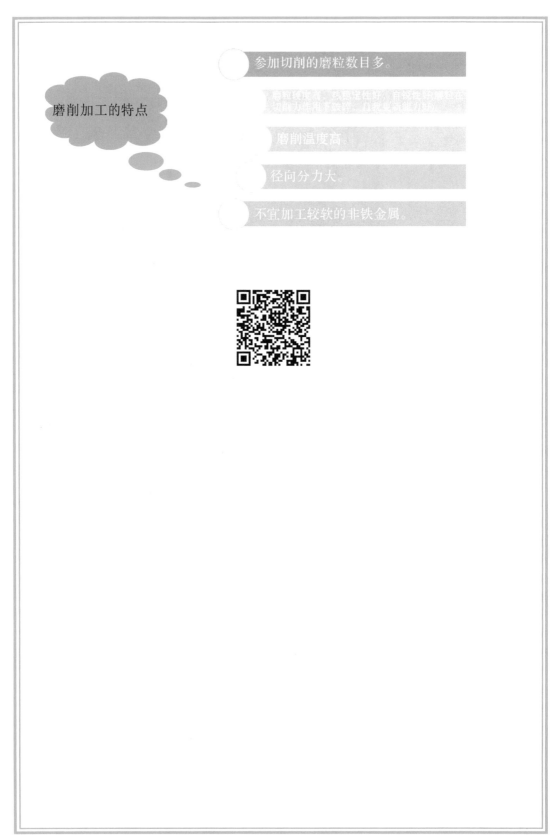

142

8.2 磨床

磨床种类繁多，本章仅对三种常用的磨床进行介绍。

外圆磨床

外圆磨床是用于加工圆柱形工件、圆锥形工件或轴肩端面和其他形状素线展成的外表面的磨床，使用非常广泛。

这几种外圆磨床是最常见的外圆磨床。端面磨床一般用于磨削工件的外圆和轴肩端面；普通外圆磨床和万能外圆磨床的区别，是万能外圆磨床不仅可以加工外圆，还可以加工内孔。

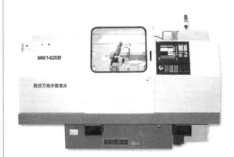
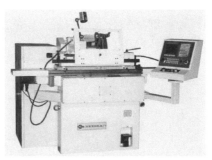

机械加工、成型和安装工艺

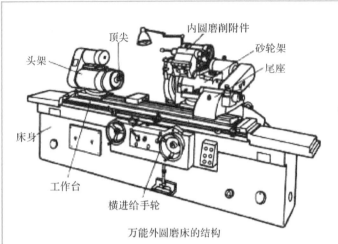

万能外圆磨床的结构

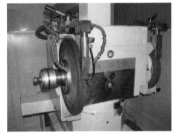 

砂轮架　　　　　　　分度头架

外圆磨床是在所有的磨床中应用最广泛的一类机床。外圆磨床部件一般都有床身、工作台、分度头架、尾座、砂轮架、进给机构和电器液压装置等。扫一扫！

  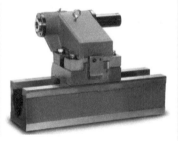

BS型万能分度头　　　顶尖　　　尾座

　　外圆磨床的主要技术参数有：最大磨削直径、最大磨削长度、最大磨削内圆直径、最小磨削内圆直径、头架顶尖孔锥度、头架主轴转速、头架回转角度、三爪自定心卡盘直径、砂轮尺寸(外径×宽度×内径)、砂轮主轴转速、砂轮架快速进给量、砂轮架横向给进量、砂轮架回转角度、工作台最大回转角度(顺时针、逆时针)、工作台纵向速度等。

内圆磨床

内圆磨床主要用于圆柱形、圆锥形内孔的磨削加工。

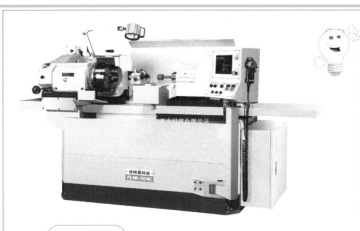

> 内圆磨床的主要技术参数有：磨孔直径、磨孔最大深度、工件最大直径（罩内、罩外）、工件最大长度、工作台最大行程、工件头架最大回转角度、砂轮最大横向移动量、砂轮转速等。

平面磨床

平面磨床是一种利用砂轮旋转加工工件，使其可以达到要求的尺寸、平面度、平行度、垂直度等公差的机床。

平面磨床主要有矩台和圆台两大类，具体又可分为卧轴矩台、卧轴圆台、立轴矩台、立轴圆台和各种专用平面磨床。

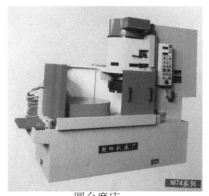
圆台磨床

矩台磨床

平面磨床的主要技术参数有：加工工件最大尺寸（长×宽×高）、砂轮主轴到工作台台面距离、工作台移动速度、砂轮尺寸（外径×宽度×孔径）等。

8.3 砂轮

砂轮是磨削加工中的主要磨具。砂轮也是特殊的刀具,其制造过程比较复杂。砂轮是将磨料和结合剂以适当比例混料成型后,再经过压坯、干燥和焙烧而制成的多孔体。

砂轮的结构包括三个要素,分别是磨料、结合剂、气孔,在三要素中磨料起到切削的作用,而结合剂把磨料连接起来并形成网状的间隙,气孔就是网状的间隙,它的作用是散热和容纳磨屑。

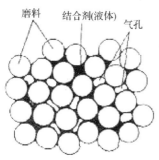

由于制造砂轮时磨料、结合剂以及制作工艺等因素影响,砂轮成形后的特性差异很大,而这些特性就是:磨料、粒度、结合剂、组织、硬度、强度、形状和尺寸等。

下面我们给大家逐一介绍!

磨料

磨具中磨粒的材料称为磨料。它是砂轮产生切削力的主要因素。因此磨料应具备硬度高,有一定的韧性、耐热性、锋利等特性。

现常用磨料分为刚玉、碳化物、立方氮化硼、金刚石几大类。

粒度

粒度是指磨料颗粒尺寸的大小。粒度有两种测量方式:筛网筛分法,沉降法。对于颗粒尺寸小于40μm的磨料,称为微粉。

结合剂

结合剂是将磨粒粘固成砂轮的材料。结合剂可分为有机结合剂和无机结合剂。粘固磨粒的结合剂有陶瓷、树脂、橡胶、菱苦土等。其中陶瓷和菱苦土为无机结合剂，树脂、橡胶为有机结合剂。

树脂结合剂由石碳酸与甲醛合成。

树脂结合剂的优点：
(1) 强度高，可制成高速砂轮和薄片砂轮；
(2) 自锐性好，砂轮不易堵塞，磨削温度低，磨削效率高；
(3) 弹性好，砂轮具有一定的弹性，可避免烧伤工件表面。同时，还具有一定的抛光作用。

树脂结合剂的缺点：
(1) 耐热性差，高温磨削下，砂轮损耗大，尤其是高负荷加工中。一般采用镀附金属磨料来解决；
(2) 化学性能不稳定，树脂容易被水、油、碱等浸蚀，故在潮湿环境中存放会降低砂轮强度，一般树脂砂轮存放保质期不超过一年。

树脂结合剂的优缺点

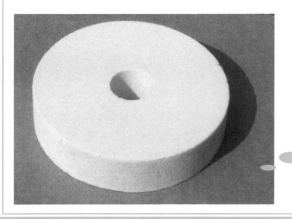

陶瓷结合剂是应用最广泛的一种结合剂，对磨料的结合强度优于树脂。一般是以天然花岗石和黏土为原料配制而成。

陶瓷结合剂的优缺点

陶瓷结合剂的优点：
(1) 力学、化学性能稳定，能耐热和耐腐蚀，适应使用各种切削液的磨削，贮存时间较长。热膨胀量小，容易控制加工精度。
(2) 散热好，由于陶瓷结合剂砂轮的多孔性，所以不易堵塞、切削锋利、磨削效率高。

陶瓷结合剂的缺点：
(1) 怕冻，陶瓷结合剂砂轮存放的时候不能处于冰冻状态，冰冻会使砂轮破裂。
(2) 易碎，不能承受大的冲击力和侧面的压力，容易使砂轮爆裂。
(3) 磨削产生的热量大。

橡胶结合剂主要原料是天然橡胶或者人造橡胶。橡胶结合剂砂轮通常气孔比较小，可制成无气孔砂轮。

橡胶结合剂的优缺点

橡胶结合剂的优点：
弹性好，可制作薄片砂轮，不易烧伤工件；抛光性能要强于树脂砂轮。

橡胶结合剂的缺点：
耐热性差，耐热温度低于150℃，易老化；橡胶耐湿性比较差，也不宜和油液接触。

硬度

在磨床加工行业中有句老话："软的磨硬的，硬的磨软的"，砂轮的硬度和我们传统意义上理解的硬度概念不一样。砂轮硬度是指磨粒在外力作用下，从结合剂中脱落的难易程度。也可以说是结合剂黏结磨粒的牢固程度。磨粒易脱落的砂轮称为软砂轮；反之，则称为硬砂轮。

砂轮的硬度直接影响砂轮的自锐性，所以一般软砂轮自锐性比较好，遇到比较硬的工件时，用软砂轮能够保持磨粒的锋利，更有利于磨削的效果。而当工件硬度比较软的时候，为了保证砂轮的使用寿命，通常会使用硬砂轮。

组织

砂轮组织是表示砂轮内部结构的松紧程度的参数，其与磨粒、结合剂、气孔三者的体积比例有关。砂轮组织的松紧程度通常用磨粒所占砂轮的百分比来表示，磨粒所占的体积百分比大，砂轮的组织就紧密，反之则组织疏松。砂轮组织关系到切削刃的多少和气孔的多少。

砂轮组织共分 15 个等级，即 0～4 级为紧密级；5～8 级为中等级；9～14 级为疏松级。

形状和尺寸

砂轮的形状是根据被加工工件来选择的，而砂轮的尺寸则是根据机床规格来选择的。

砂轮的强度

砂轮的强度一般指砂轮的最高工作速度。

砂轮高速旋转时，砂轮上任意一部分都受到很大的离心力作用，如果砂轮没有足够的回转强度，砂轮就会爆裂而引起严重的事故。所以安装砂轮前一定要检查砂轮标识上的砂轮工作最高线速度是多少，一定要小于机床工作线速度。一般国际上有一个安全指标的规定，砂轮实际最高工作线速度是在乘以安全系数 $f \approx 1.25$~1.5 以后的线速度，而不是标识上的线速度，但安装砂轮一定要低于机床线速度。

8.4 砂轮的安装和平衡

砂轮的安装

用法兰盘装夹砂轮时，在安装前应仔细检查是否有裂缝。检查的方法是用绳子穿入砂轮内孔将砂轮吊起来，用木锤子轻轻敲击砂轮，如果声音清脆说明砂轮没有裂纹。砂轮的孔径与法兰盘的配合应作间隙配合，以避免在加工过程中法兰受热膨胀使砂轮胀裂。法兰盘大小应该为砂轮最大直径的三分之一左右，以便于分散压紧力，不会将砂轮压裂。在砂轮和法兰盘之间应放用弹性材料制成的衬垫，一般为软纸板、毛毡等。衬垫的厚度在 0.5～1mm 之间。紧固砂轮法兰盘时，螺母不要拧得太紧，螺钉不可拧紧，应按对称位置依次拧紧。拧紧时，只能用标准扳手，不能用敲打的方法加大拧紧力，也不能用加力杆。

砂轮的平衡调整

(1) 把砂轮安装好并擦拭锥孔，用件 2 平衡芯轴穿过件 3 法兰盘安装好，将法兰盘侧壁内的平衡块全部拆下，清洗干净。

(2) 用水平仪校平平衡架导轨，把装好的砂轮放在件 5 导轨上，安放时要求芯轴与导轨垂直，慢慢转动砂轮，砂轮停止时则砂轮重心在下，在重心对面做一个记号。

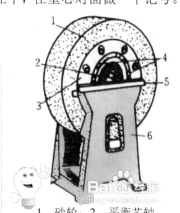

(1) 平衡块一共有三块，在重心一边装一块平衡块，使记号位置不变，然后在记号两边各放一块，调整 2 和 3 处两块平衡块的位置；保持记号位置不变。

(2) 把砂轮旋转 90°，观察砂轮平衡状态，如不平衡调整 2 和 3 处两块平衡块到平衡为止。

(3) 把砂轮旋转 180°，按上述方法调整至平衡，使砂轮在所有位置不动。砂轮粗平衡调整完毕。

(4) 将粗调整完的砂轮安装到机床上，用金刚石笔修整至圆，然后拆卸砂轮，按上述方法进行精平衡调整。

1—砂轮；2—平衡芯轴；
3—法兰盘；4—平衡块
5—导轨；6—支架
砂轮的结构

砂轮修整

修整砂轮用工具：金刚石笔，碳化硅砂块，成形滚轮等。

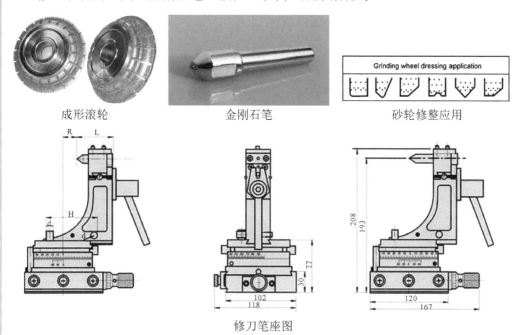

成形滚轮　　　　金刚石笔　　　　砂轮修整应用

修刀笔座图

修整砂轮常用夹具和附件：修整杆或修整座、角度修正弯板、滚轮架和四连杆靠模仿形器等。

修整方法如下：

(1) 砂轮端面各种型面的粗修，一般是用手紧握砂块，操作者站在砂轮旋转方向的侧面，按照加工的要求进行修整。

(2) 砂轮端面及各型面的精修，则是用金刚石笔或滚轮安装在各种夹具或附件上进行。金刚石笔一般 10°～15°安装，修整用量要合理，并应充分冷却。

8.5 常见磨削机床的加工方法

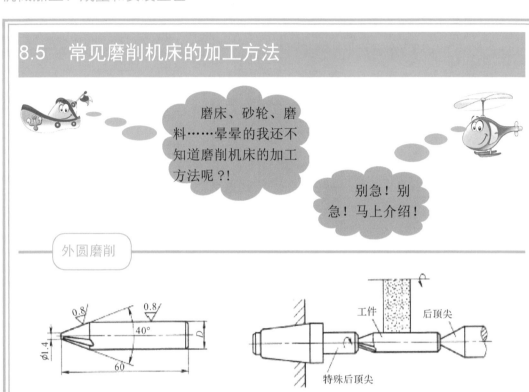

外圆磨削

(a) 工件

(b) 双顶尖装夹磨削工件外圆

外圆磨床一般用于回转体外圆面及外锥面的加工。工件的装夹常用双顶尖装夹，装夹前要检查中心孔是否合格。装夹力要适中。

用卡盘装夹工件前卡盘需要校正。

外圆磨床常用的磨削方式

1. 横向磨削法

横向磨削法也称切入磨削法，适合加工短而粗及带台阶的轴类工件的外圆。此外，成形磨削也采用此种方法。

横向磨削法砂轮只作横向进给运动，不作纵向运动。横向磨削法效率高，但是磨削时不容易散热，而且砂轮表征容易反映在工件上面。

2. 纵向磨削法

纵向磨削法应用广泛，适用于加工可以纵向走刀的各种轴类工件的外圆。

纵向磨削法先切入工件，然后砂轮作纵向进给磨削工件，砂轮回程时不作横向进给。

纵向磨削法加工产生的热量低，工件成形精度高，表面粗糙度好，但是加工效率比较低。

3. 阶梯磨削法

此方法是一种高效的磨削方法,适用于余量较大、刚性好的批量生产的工件。阶梯磨削法的磨削步骤与纵向磨削法一致。但是,由于磨削面宽、吃刀量大,工件易发热变形,故必须注意充分冷却。

我就说不好理解嘛!什么纵向横向的!哎……

内圆磨削

内圆磨削一般用于加工轴类或盘类工件的内孔。

工件的装夹方法如下:

工件一般用三爪自定心卡盘或四爪单动卡盘装夹,工件装夹后,要用铜棒敲击找正,敲击过程中要用表压住工件,来检测工件是否找正,敲击时一定要抬起表头,以防表损坏。工件过长的时候需要用中心架来辅助支撑。

中心架

磨削方法及注意事项:

调整砂轮的行程,工件两端需越程,越程长度约为工件的 $1/3$ 到 $1/2$ 间,以防止工件两端产生喇叭口。加工完毕如产生锥度,适当调整机床角度修整工件,直至符合要求。磨削内孔一般采用纵向磨削法,接长轴尽量加粗,以保证刚度。

平面磨削

平面磨削用于加工平面工件,以保证工件的平面度、平行度、垂直度等公差。

平面磨床的磨削方法:

平面磨床有很多种类,这里介绍矩台卧轴平面磨床的加工方法。

(1) 横向磨削法。当工件长而宽的时候,每当工作台纵向行程终了时,砂轮主轴作一次横向进给,待工件上去除一层金属后,砂轮再作垂直进给,直至切去全部余量。

(2) 切入磨削法。磨削时,砂轮不作横向进给,只是在磨削将要结束时,作适当的横向移动(适用于工件磨削面宽度小于砂轮宽度的情况)。

平面磨削工件的装夹方法:

装夹前,检查磁力吸盘的平面度和平行度,能否达到工件加工的技术要求。

(1) 工件直接装夹在磁力吸盘上。装夹时应注意的事项如下:

① 装夹前必须先擦拭磁力吸盘和工件，用油石去除毛刺。

② 工件应尽可能多地压在磁力吸盘的磁力线上以保证工件吸牢，有利于磨削加工。

③ 加工平行面时，如果工件形状不规则，应先以大面为基准磨小面，以后再互为基准交替磨削至尺寸。

④ 在以小面为基准磨工件的大面时，工件受切削力影响的两端应加挡铁，挡铁的基面应大而平直，高度至少不低于工件高度的三分之二。

⑤ 工件批量磨削时，在矩形磁力吸盘上装夹工件应基本上平行于工作台导轨，纵向成一条线。

(2) 当加工一些非金属材料，如塑料、陶瓷等不导磁工件时，或加工一些工件的垂直面时，需其他夹具辅助装夹。一般常用精密虎钳，虎钳使用前需拉平找正。

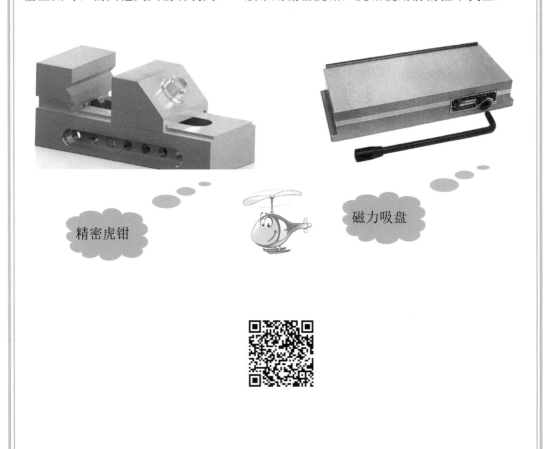

精密虎钳　　　磁力吸盘

第九章 特种加工

9.1 概述

 为什么要学习特种加工呢？它很重要吗？

自20世纪50年代以来，由于机械、电子和航空航天等工业的迅猛发展，各种新材料大量涌现，精密零件的形状更趋复杂。这些高新技术的发展对机械加工的要求日益提高，这都给机械加工领域带来前所未有的挑战，具体如下：

◆ 大量采用硬、脆、软等新型材料（如超硬合金、工程陶瓷、复合材料）的零件；

◆ 具有高精度复杂机械结构的零件（如喷气发动机涡轮叶片、非球面镜、军工光学部件）；

◆ 尺寸特大或微小零件的加工（如核电风电设备、半导体芯片制造、航天发动机喷嘴）；

◆ 采用薄壁低刚度结构的零部件加工（如航天器保护罩）。

传统的切削加工工艺方法难以承担以上所述的零件加工。目前，人们找到了一种特殊的加工方法，并成为机械制造业中不可缺少的工艺手段——特种加工。

特种加工

特种加工是指不属于传统加工工艺范畴的加工方法，它不同于使用刀具、磨具等直接利用机械能切除多余材料的传统加工方法，而是利用电能、热能、声能、光能、化学能和电化学能等，有时也结合机械能对工件进行的加工。

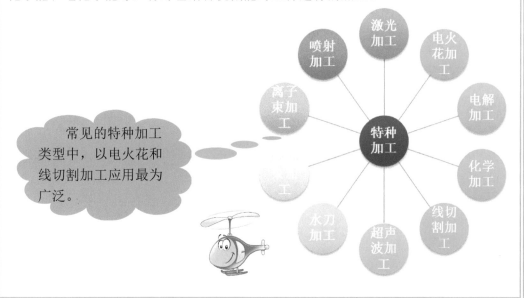

常见的特种加工类型中，以电火花和线切割加工应用最为广泛。

机械加工、成型和安装工艺

激光加工

激光加工是将从激光器发射出的激光，经光路系统，聚焦成高功率密度的激光束。激光束照射到工件表面，使工件达到熔点或沸点，同时与光束同轴的高压气体将熔化或汽化金属吹走。随着光束与工件相对位置的移动，最终使材料形成切缝，从而达到切割的目的。

激光切割机加工中

激光切割的零件展示

水刀加工

水刀又称水加工，即高压水射流切割技术，是一种利用高压水流切割的机器。在电脑的控制下能任意雕琢工件，而且受材料质地影响小。因为成本低，易操作，成品率又高，水切割正逐渐成为工业切割技术方面的主流切割方式。

水刀切割中

水刀切割机床

超声波加工

超声波加工是利用超声波频率作小振幅振动的工具，通过它与工件之间游离于液体中的磨料对被加工表面的捶击作用，使工件材料表面逐步破碎的特种加工，英文简称为 USM。超声波加工常用于穿孔、切割、焊接、套料和抛光。

超声波焊接机

超声波抛光机

看了这么多的图片，介绍了这么多的加工方式，我们还是重点说说电火花和线切割加工吧！

在介绍之前，我先考一考你吧！你知道什么是"电蚀现象"吗？

机械加工、成型和安装工艺

什么是"电蚀现象"？

我们称"电蚀现象"为蚀除原理，当两电极间的距离足够小时，电极间的绝缘介质（气体、液体）被极化击穿形成放电通道，放电通道内瞬间产生极大的热量，足以使电极材料表面局部熔化或汽化，并在一定条件下，熔化或汽化的部分脱离电极表面，形成放电蚀穴，这种现象叫做电蚀现象。

所有的电加工设备使用的加工原理，都是利用"电蚀现象"进行金属材料的去除。

Special processing

It refers to the traditional processing methods which do not belong to the traditional processing technology. It is different from the traditional processing methods which use cutting tools, abrasives and other direct use of mechanical energy to remove redundant materials. It directly uses electric energy, thermal energy, sound energy, light energy, chemical energy and electrochemical energy, and sometimes also combines mechanical energy to process the workpiece.

laser machining

Laser cutting is to focus the laser beam with high power density through the optical system emitted from the laser. The laser beam irradiates the surface of the workpiece, which makes the workpiece reach the melting point or boiling point. At the same time, the high-pressure gas coaxial with the beam will blow away the melted or gasified metal. With the relative position of the beam and the workpiece moving, the material will eventually form a slit to achieve the purpose of cutting.

Water jet machining

Water cutting, also known as water knife, is a high-pressure water jet cutting technology as well as a high-pressure water cutting machine. Under the control of computer, the workpiece can be carved arbitrarily, and it is less affected by the material quality. Low in cost, easy in operation and high in production, water cutting is becoming the mainstream cutting method in industrial cutting technology.

Ultrasonic machining

Ultrasound machining (USM) is a kind of special processing, which uses ultrasonic frequency as a tool for small amplitude vibration, and through the impact of abrasive free from the liquid between the tool and the workpiece on the machined surface, the surface of workpiece material is gradually broken up. Ultrasound machining is abbreviated as USM. Ultrasound processing is often used in perforation, cutting, welding, blanking and polishing.

What is "erosion phenomenon"?

When the distance between the two electrodes is small enough, the insulating medium (gas and liquid) between the electrodes is polarized to break down and form a discharge channel. The discharge channel instantaneously generates a great deal of heat, which is enough to make the surface of the electrode material partial melt or vaporize, and under certain conditions, melt or vaporize. The electrochemical etching phenomenon is called electrochemical etching phenomenon, in which the part of the electrochemical etching is separated from the surface of the electrode and forms discharge cavities.

9.2 电火花加工

早在十九世纪，人们就发现了电器开关的触点开闭瞬间会打火放电，使接触部位烧蚀，造成接触面的损坏，这种放电引起的电极烧蚀现象叫做电腐蚀。为了减少和避免这种有害的电腐蚀，人们一直在研究电腐蚀产生的原因和防止的方法。当人们掌握了它的规律之后，便创造条件，转害为益，把电腐蚀用于生产中，从而衍生出电蚀现象和电加工工艺。

◆ 1943 年，前苏联科学院院士拉扎连柯研制出 RC 放电回路的电火花加工装置；

◆ 20 世纪 50 年代，西班牙欧纳、日本三菱、瑞士阿奇公司研制出电火花加工机床；

◆ 1951 年起，我国开始了电火花加工的实验研究；

◆ 1959 年，航空机电、哈工大、电工所、上交大等单位派人到前苏联学习电加工设备制造及工艺；

◆ 1963 年，我国研制出高速走丝数控线切割机床样机。

电火花的前世今生

不要惊叹不已啦，下面来一个烧脑的内容，介绍一下吧！

电火花加工原理

电火花加工时，脉冲电源的一极接工具电极，另一极接工件电极，两极均浸入具有绝缘度的液体介质（常用煤油、矿物油或去离子水）中。工具电极由自动进给调节，以保证工具与工件在正常加工时维持一很小的放电间隙（0.01～0.05 mm）。当脉冲电压加到两极之间，便将当时条件下极间最近点的液体介质击穿，形成放电通道。由于通道的截面积很小，放电时间极短（10^{-6} ～ 10^{-9} s），致使能量高度集中（10～107 W/mm），放电区域产生的瞬间高温足以使材料熔化甚至蒸发，由于熔化和汽化的速度很高，故带有爆炸性。在爆炸力的作用下，将熔化了的金属微粒迅速抛出，被液体冷却、凝固后从间隙中冲走，以致形成一个小凹坑。第一次放电结束之后，经过很短的间隔时间，第二个脉冲又在另一极间最近点击穿放电。如此周而复始高频率地循环下去，工具电极不断地向工件进给，它的形状最终就复制在工件上，形成所需要的加工表面。与此同时，总能量的一小部分也释放到工具电极上，从而造成工具损耗。

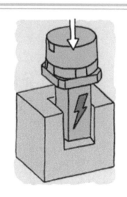

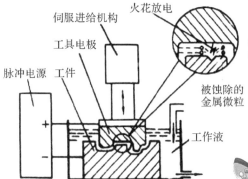

电火花加工原理图

我有扫一扫！我怕谁？！

电火花的加工特点

- 脉冲放电的能量密度高，用传统机械加工方法难以加工或无法加工的特殊材料、复杂结构的工件可使用电火花加工方法。

- 脉冲放电持续时间极短，放电时产生的热量传导扩散范围小，加工表面的材料受热的影响小。

- 加工时，工具电极与工件材料不接触，两者之间宏观作用力小。所以，工具电极材料无须比工件材料硬，因此，工具电极容易制造。

- 直接利用电能进行加工，便于实现加工过程的自动化，并可减少机械加工工序，加工周期短、劳动强度低、使用维护方便。

电火花加工的条件

(1) 工具电极和工件之间必须维持一定的放电间隙，放电间隙的大小既可以满足脉冲电压不断击穿介质产生火花放电，又可以在放电通道关闭后介质消电离，以排除蚀除产物的要求。

(2) 在放电加工中，两电极之间必须充入绝缘介质（液体）。

(3) 输送到两电极间的脉冲能量密度应足够大（$10^5 \sim 10^6 \text{A/cm}^2$），在放电通道形成后，脉冲电压的变化足够小，并必须维持足够大的峰值电流，通道才可以在高电平期间得到维持。一般情况下，维持通道的峰值电流不小于2A。

(4) 加工中必须使用脉冲放电，放电持续时间（高电平）一般为 $10^{-6} \sim 10^{-9}$s，由于放电时间很短，使放电时产生的热能来不及在被加工材料内部扩散，从而把能量作用在局部很小范围内，保持加工过程的冷极特性。

(5) 脉冲放电需要重复多次进行，并且多次脉冲放电在时间上和空间上是分散的，即每次脉冲放电一般不在同一点进行，避免发生局部烧伤。

(6) 脉冲放电后的电蚀产物能及时排运至放电间隙之外，使重复性脉冲放电顺利进行。

电火花加工的主要用途

(1) 加工各种金属及其合金，特殊的热敏材料、半导体和非导体材料。

(2) 加工各种复杂形状的型孔，包括圆孔、方孔、多变孔、异形孔、曲线孔、螺纹孔、微孔、深孔等，以及各种型面的型腔，如数微米的孔、槽，数米的超大型模具和零件。

(3) 各种材料及工件的切割，包括材料的切断、特种结构件的切割、切割微细窄缝组成的零件（如金属栅网、异形孔喷丝板、激光器件等）。

(4) 加工各种刀具、样板、工具、量具、螺纹等成型零件。

(5) 工件的磨削，包括小孔、深孔、内圆、外圆、平面等成型磨削。

(6) 刻写、打印铭牌、标志等。

(7) 表面强化，如金属的表面高频淬火、渗碳、涂覆特殊材料及其合金化等。

(8) 一些辅助用途，如去除断在零件中的丝锥、钻头，修复磨损件，跑合齿轮啮合件等。

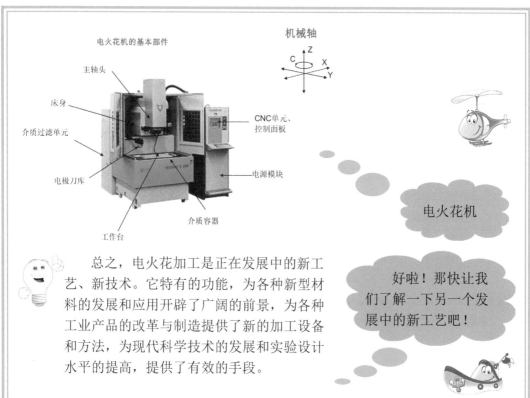

总之，电火花加工是正在发展中的新工艺、新技术。它特有的功能，为各种新型材料的发展和应用开辟了广阔的前景，为各种工业产品的改革与制造提供了新的加工设备和方法，为现代科学技术的发展和实验设计水平的提高，提供了有效的手段。

Principle of EDM

In EDM, one pole of pulse power supply is connected with tool electrode and the other pole is connected with workpiece electrode. Both poles are immersed in insulating liquid medium (kerosene or mineral oil or deionized water). Tool electrodes are adjusted by automatic feeding to ensure a small discharge gap (0.01-0.05 mm) between tool and workpiece during normal processing. When the pulse voltage is applied between the two poles, the liquid dielectrics at the nearest point between the poles will break down and form a discharge channel. Because of the small cross-section area and short discharge time (10^{-6}-10^{-9}s), the energy is highly concentrated (10-107W/mm). The instantaneous high temperature generated in the discharge area is enough to melt or even evaporate the material. Because of the high speed of melting and vaporization, it is explosive. Under the action of explosive force, molten metal particles are rapidly thrown out, cooled, solidified by liquid and washed away from the gap, thus forming a small pit. At the end of the first discharge, and after a short interval, the second pulse clicks through the discharge nearest between the other poles. The tool electrodes are continuously fed to the workpiece, and the shape of the tool electrodes is eventually replicated on the workpiece to form the required machined surface. At the same time, a small part of the total capacity is released to the tool electrodes, resulting in tool wear.

EDM Processing Conditions

1. A certain discharge gap must be maintained between tool electrodes and workpieces. The size of discharge gap can not only satisfy the requirement of continuous breakdown of

dielectrics by pulse voltage and spark discharge, but also deionize dielectrics after the discharge channel is closed to eliminate the requirement of corrosion products.

2. In EDM, the insulating medium (liquid) must be filled between the two electrodes.

3. The pulse energy density between the two electrodes should be large enough (10^5-$10^6 A/cm^2$). After the discharge channel is formed, the change of the pulse voltage is small enough and the peak current must be maintained sufficiently large for the channel to be maintained during the high level period. In general, the peak current of the maintenance channel is not less than 2A.

4. Pulse discharge must be used in machining. The duration of discharge (high level) is generally $10^{-6} \sim 10^{-9}$S. Because the discharge time is very short, the thermal energy generated during discharge can not diffuse in the material being processed, so that the energy can be applied in a very small range to maintain the cold pole characteristics of the machining process.

5. Pulse discharges need to be repeated many times, and multiple pulse discharges are dispersed in time and space, that is, each pulse discharges are generally not carried out at the same point to avoid local burns.

6. The corrosion products after pulse discharge can be discharged out of the discharge gap in time to make the repetitive pulse discharge go smoothly.

Main Uses of EDM

1. Processing various metals and their alloys, special thermal sensitive materials, semiconductors and non-conductors.

2. Processing various complex shaped holes, including processing round holes, square holes, multi-variable holes, special-shaped holes, curve holes, thread holes, micro-holes, deep holes and other types of cavity, such as micron holes, grooves, several meters of super-large die and parts.

3. Cutting of various materials and workpieces, including cutting of materials, cutting of special structural components, cutting parts consisting of tiny slits (such as metal mesh, special hole spinneret, laser parts, etc.).

4. Processing various forming parts such as cutters, templates, tools, measuring tools, threads, etc.

5. Workpiece grinding, including forming grinding of small holes, deep holes, inner circle, outer circle, plane, etc.

6. Write, print nameplates, logos, etc.

7. Surface strengthening, such as high frequency quenching, carburizing, coating special materials and alloying of metal surface.

8. Some auxiliary uses, such as removing broken taps, drills, repairing wear parts, running-in gear meshing parts, etc.

9.3 线切割加工

金属导线

与电火花机床相同,线切割机床也是利用电蚀现象对导电材料进行切割的,但是不同的是线切割机床采用金属导线(电极丝)代替了电火花机床的实体电极材料,电极丝按照编制好的运行轨迹在工件表面进行 X/Y 方向进给,从而切割出编程轮廓。切割过程中电极丝会参与放电,切削工件,但同时自身面向行进方向的半个直径表面也会腐蚀损耗,为了提高加工精度和表面质量,慢走丝线切割的电极丝被设计成一次性使用。

在慢走丝线切割中,电极丝通常由黄铜制成并且直径为 0.25 ± 0.002 mm,所使用的介质是在合适的压力下注入切割区域的去离子水。

放电区域的关键部件

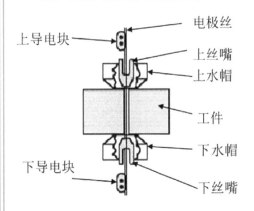

线切割电极丝的引导、固定方式（如左图），上下丝嘴（孔径 $0.25^{+0.005}$ mm）负责固定电极丝在 X/Y 平面的正确位置；上下导电块负责将电极丝连通到电源模块的电路中；上下水帽负责聚拢去离子水压，形成良好的冷却及循环目的。

线切割加工的特点

(1) 线切割加工是轮廓切割加工，无需设计和制造成形工具电极，大大降低了加工费用，缩短了生产周期。

(2) 直接利用电能进行脉冲放电加工，工具电极和工件不直接接触，无机械加工中的宏观切削力，适宜于加工低刚度零件及细小零件。

(3) 无论工件硬度如何，只要是导电或半导电的材料都能进行加工。

(4) 切缝可窄达 0.005 mm，只对工件材料沿轮廓进行"套料"加工，材料利用率高，能有效节约贵重材料。

(5) 移动的长电极丝连续不断地通过切割区，单位长度电极丝的损耗量较小，加工精度高。

(6) 一般采用水基工作液，可避免发生火灾，安全可靠，可实现昼夜无人值守连续加工。

(7) 通常用于加工零件上的直壁曲面，通过 X-Y-U-V 四轴联动控制，也可进行锥度切割和加工上下截面异形体、形状扭曲的曲面体和球形体等零件。

(8) 不能加工盲孔及纵向阶梯表面。

实际加工中，线切割机床还会分为快走丝（高速走丝）线切割和慢走丝（低速走丝）线切割，两种机床各有优缺点，让我们通过下表了解一下吧！

比较项目	高速走丝线切割加工	低速走丝线切割加工
走丝速度	8～10 m/s	0.001～0.25 m/s
走丝方式	往复供丝，反复使用	单向走丝，一次性使用
电极丝材料	钼、钨钼合金	黄铜丝，以铜为主体的镀覆材料
穿丝方式	手动	手动或自动
电极丝振动	较大	较小
运丝系统结构	简单	复杂
工作液	乳化液	去离子水或煤油
工作液电阻率	0.5～50 kΩ·cm	10～100 kΩ·cm
导丝机构形式	导轮，寿命短	导向器，寿命长
机床价格	便宜	昂贵
切割速度	20～160 mm^2/min	20～240 mm^2/min
加工精度	0.01～0.04 mm	0.004～0.01 mm
表面粗糙度Ra	1.6～3.2 μm	0.1～1.6 μm
重复定位精度	0.02 mm	0.004 mm
电极丝损耗	均匀损耗	不计损耗

通过上面的对比，我们发现快走丝线切割的主要优点是机床造价和运行成本都比慢走丝线切割低廉很多，但是切割精度低、表面质量不高的缺点也很明显。所以在实际加工中要根据工件的实际要求来进行有针对性的选择使用。

The same as EDM machine tools, WEDM machine tools cut conductive materials by means of electrical erosion, but the difference is that WEDM machine tools use metal wires (electrode wires) instead of solid electrode materials of EDM machine tools. The electrode wires are fed in the X Y direction on the workpiece surface according to the prepared trajectory. Cut out the programming outline. In order to improve the machining accuracy and surface quality, WEDM wire is designed to be used in one time.

In WEDM, the electrode wire is usually made of brass with a diameter of $0.25^{+0.005}$ mm. The medium used is deionized water injected into the cutting area under suitable pressure.

Characteristics of WEDM

1. WEDM is a contour cutting process. It does not need to design and manufacture tool electrodes, which greatly reduces the processing costs and shortens the production cycle.

2. Direct use of electrical energy for pulse discharge machining, tool electrodes and workpieces do not contact directly, there is no macro cutting force in mechanical processing, suitable for processing low rigidity parts and small parts.

3. No matter how hard the workpiece is, as long as it is conductive or semi-conductive material, it can be processed.

4. The slit can be narrowed to only 0.005 mm, and only the material of the workpiece is processed along the contour. The material utilization rate is high and the precious material can be effectively saved.

5. The moving long electrode wire passes through the cutting area continuously. The loss per unit length of the electrode wire is small and the processing accuracy is high.

6. Usually water-based working fluid is used to avoid fire, which is safe and reliable, and can realize continuous processing without person on duty day and night.

7. Usually used for processing straight-wall surfaces on parts, through X-Y-U-V four-axis linkage control, taper cutting and processing of upper and lower cross-section profiles, distorted curved surfaces and spheres and other parts.

8. Blind holes and longitudinal stepped surfaces cannot be machined.

第四篇　机械加工工艺规程与车间管理

了解了金属材料的性能及多种加工工艺方法后,如何利用这些方法加工我们所需要的零件呢?如何提高加工效率?生产环境对加工又起哪些作用呢?本篇将介绍机械加工工艺规程设计、车间布局及5S管理。

第十章 机械加工过程与工艺规程

10.1 概述

在之前的学习中，我们已经了解了成型加工、机械加工、特种加工等多种加工方式。在实际的生产过程中，如何根据实际加工要求，合理编排这些加工方式，正确运用这些加工方式呢？

本章介绍机械加工工艺规程的设计及机械加工过程的编排。

10.2 基本概念

生产过程

机械产品的生产过程是指将原材料转变为成品的所有劳动过程。对于机械制造而言，生产过程包括：

(1) 原材料、半成品和成品的运输和保存。

(2) 生产和技术准备工作，如产品的开发和设计、工艺及工艺装备的设计与制造、各种生产资料的准备以及生产组织。

(3) 毛坯制造和处理。

(4) 零件的机械加工、热处理及其他表面处理。

(5) 部件或产品的装配、检验、调试、油漆和包装等。

工艺过程

工艺过程是指在生产过程中，直接改变生产对象的形状、尺寸、相对位置或性能，使之成为成品或半成品的过程。

工艺过程可分为毛坯制造、机械加工、热处理和装配等。

◆机械加工工艺过程是指用机械加工的方法直接改变毛坯的形状、尺寸和表面质量，使之成为零件或部件的那部分生产过程。

在机械加工工艺过程中，根据零件的结构特点和技术要求，要运用不同的加工方法、设备和刀具等，同时还需要按照一定的顺序进行加工，才能完成由毛坯到零件的过程。

机械加工、成型和安装工艺

◆装配工艺过程是把零件装配成机械并达到装配要求的过程。

工 序

工序是指一个或一组工人,在一个工作地点对同一个或同时对几个工件进行加工所连续完成的那部分工艺过程。

判别是否为同一工序的主要依据是:工作地点是否变动和加工是否连续。

安 装

安装是指将工件在机床或夹具中定位、夹紧一次所完成的那一部分工序内容。

◆定位:在加工前,应先使工件在机床上或夹具中占有正确的位置,这一过程称为定位。

◆夹紧:工件定位后,将其固定,使其在加工过程中保持定位位置不变的操作过程,称为夹紧。

工 位

工位是指为了完成一定的工序内容,一次安装工件后,工件与夹具或设备的可动部分一起,相对刀具或设备的固定部分所占据的每一个位置。

◆多工位加工:为了减少由于多次安装带来的误差和时间损失,加工中常采用回转工作台、回转夹具或移动夹具,使工件在一次安装中,先后处于几个不同的位置进行加工。

工步

工序又可分成若干工步。加工表面不变、切削刀具不变、切削用量中的进给量和切削速度基本保持不变的情况下连续完成的那部分工序内容，称为工步。

三个不变因素中只要有一个因素改变，即成为新的工步。

一道工序包括一个或几个工步。

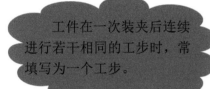

工件在一次装夹后连续进行若干相同的工步时，常填写为一个工步。

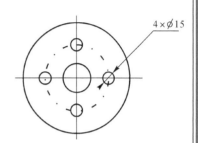

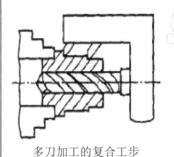

多刀加工的复合工步

用几把刀具或复合刀具，同时加工同一工件上的几个表面，称为复合工步。在工艺文件上，复合工步视为一个工步。

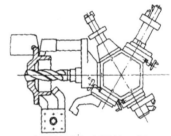

立轴转塔车床回转刀架

进给

刀具从被加工表面每切下一层金属即称为一次进给。一个工步可能只要一次进给，也可能要几次进给。

生产纲领

工厂在计划期内应当生产的产品产量和进度计划，称为生产纲领。工厂一年制造的合格产品的数量，称为年生产纲领，也称年产量。产品中某零件的生产纲领除计划规定的数量外，还必须包括备品率及平均废品率，即：

$$N = Qn(1 + \alpha) \cdot (1 + \beta)$$

式中：N——零件的年产量(件/年)；

Q——产品的年产量(台/年)；

n——每台产品中，该零件的数量(件/台)；

α——备品率，以百分数表示；

β——废品率，以百分数表示。

生产类型

生产类型是对企业生产专业化程度的分类。根据产品大小和生产纲领的不同，一般把机械制造生产分为单件生产、成批生产和大量生产三种类型。

生产类型和生产纲领的关系

生产类型		同种零件的年产量/件		
		重型（30 kg以上）	中型（4～30 kg）	轻型（4 kg以下）
单件生产		5以下	10以下	100以下
成批生产	小批生产	5～100	10～200	100～500
	中批生产	100～300	200～500	500～5000
	大批生产	300～1000	500～5000	5000～50000
大量生产		1000以上	5000以上	50000以上

各种生产类型的工艺特征

项目	单件、小批生产	成批生产	大批、大量生产
产品数量	少	中等	大量
加工对象	经常变换	周期性变换	固定不变
毛坯制造	手工造型和自由锻	部分采用金属模造型和模锻	机器造型、压力铸造、模锻
设备和布置	通用设备（万能的），按机群布置	通用的和部分专用设备，按工艺路线布置成流水线	广泛采用高效率专用设备和自动化生产线
夹具	通用夹具	广泛使用专用夹具和特种工具	广泛使用高效率专用夹具和特种工具
刀具和量具	一般刀具、通用夹具和量具	部分采用专用刀具和量具	高效率专用刀具和量具
安装方法	划线找正	部分划线找正	不需划线找正
加工方法	根据测量进行试切	用调整法加工，可组织成组加工	使用调整法自动加工
装配方法	钳工试配	普遍应用互换性，保留某些试配	全部互换，不需钳工试配
工人技术水平	需技术熟练	需技术比较熟练	技术熟练程度要求低
生产率	低	中	高
成本	高	中	低
工艺文件	编写简单工艺过程卡	详细编写工艺卡	详细编写工艺卡和工序卡

机械加工工艺规程

机械加工工艺规程是将产品或零部件的制造工艺过程和操作方法按一定格式固定下来的技术文件。它是在具体生产条件下，本着最合理、最经济的原则编制而成，经审批后用来指导生产的法规性文件。

机械加工工艺规程是机械制造工厂最主要的技术文件，是工厂规章制度的重要组成部分，其作用主要有：

- 它是组织和管理生产的基本依据。
- 它是指导生产的主要技术文件。
- 它是新建和扩建工厂的原始资料。
- 它是进行技术交流，开展技术革新的基本资料。

机械加工工艺规程包括：零件加工工艺流程、加工工序内容、切削用量、采用设备及工艺装备、工时定额。

根据原机械电子工业部指导性技术文件 JB/Z338.5《工艺管理导则 工艺规程设计》中规定，工艺规程的类型有：

◆专用工艺规程——针对每一个产品和零件所设计的工艺规程；

◆通用工艺规程，它包括以下三种：

(1) 典型工艺规程——为一组结构相似的零部件所设计的通用工艺规程；
(2) 成组工艺规程——按成组技术原理将零件分类成组，针对每一组零件所设计的通用工艺规程；
标准工艺规程——已纳入国家标准或工厂标准的工艺规程。

为了适应工业发展的需要，加强科学管理和便于交流，原机械电子工业部还制定了指导性技术文件 JB/Z 187.3—88《工艺规程格式》，按照规定，属于机械加工工艺规程的有：

(1) 机械加工工艺过程卡片：主要列出零件加工所经过的整个工艺路线，以及工装设备和工时等内容，多作为生产管理使用；

(2) 机械加工工序卡片：用来具体指导工人操作的一种最详细的工艺文件，卡片上要画出工序简图，注明该工序的加工表面及应达到的尺寸精度和粗糙度要求、工件的安装方式、切削用量、工装设备等内容；

(3) 标准零件或典型零件工艺过程卡片；

(4) 单轴自动车床调整卡片；

(5) 多轴自动车床调整卡片；
(6) 机械加工工序操作指导卡片；
(7) 检验卡片。

属于装配工艺规程的有：
(1) 装配工艺过程卡片；
(2) 装配工序卡片。

工艺规程的要求：
- 必须充分利用本企业现有的生产条件。
- 必须可靠地加工出符合图纸要求的零件，保证产品质量。
- 保证良好的劳动条件，提高劳动生产率。
- 在保证产品质量的前提下，尽可能降低消耗、降低成本。
- 应尽可能采用国内外先进工艺技术。

制订工艺规程的主要依据

- 产品的装配图和零件的工作图
- 产品的生产纲领
- 本企业现有的生产条件，包括毛坯的生产条件或协作关系、工艺装备和专用设备及其制造能力、工人的技术水平以及各种工艺资料和标准等
- 产品验收的质量标准
- 国内外同类产品的新技术、新工艺及其发展前景等的相关信息

制定机械加工工艺规程的步骤大致如下：
(1) 熟悉和分析制定工艺规程的主要依据，确定零件的生产纲领和生产类型；
(2) 分析零件的工作图和产品装配图，进行零件结构工艺性分析；
(3) 确定毛坯，包括选择毛坯类型及其制造方法；
(4) 选择定位基准或定位基面；
(5) 拟订工艺路线；
(6) 确定各工序需用的设备及工艺装备；

(7) 确定工序余量、工序尺寸及其公差；
(8) 确定各主要工序的技术要求及检验方法；
(9) 确定各工序的切削用量和时间定额，并进行技术经济分析，选择最佳工艺方案；
(10) 填写工艺文件。

制定工艺规程时，主要解决以下几个问题。

- 零件图的研究和工艺分析。
- 毛坯的选择。
- 定位基准的选择。
- 工艺路线的拟订。
- 工序内容的设计，包括机床设备及工艺装备的选择、加工余量和工序尺寸的确定、切削用量的确定、热处理工序的安排、工时定额的确定等。

Productive Process

The production process of mechanical products refers to all the labor processes that transform raw materials into finished products.

For mechanical manufacturing, the production process includes:

1. Transportation and preservation of raw materials, semi-finished products and finished products.

2. Production and technical preparation, such as product development and design, process and equipment design and manufacture. Preparation of various means of production and production organization.

3. Making and processing of hair germ.

4. Machining, heat treatment and other surface treatment of parts.

5. Assembly, inspection, commissioning, painting and packaging of components or products.

Process

Process refers to the process of directly changing the shape, size, relative position or performance of the production object into finished or semi-finished products in the production process.

Machining process refers to the process of directly changing the shape, size and surface quality of hair germ by means of mechanical processing, which is called part or part production process.

In the process of mechanical processing, according to the structural characteristics and technical requirements of parts, different processing methods, equipment and cutting tools should be used. At the same time, a certain sequence of processing is needed to complete the process from hair germ to parts.

Assembly process is the process of assembling parts into machinery and meeting assembly requirements.

Process

Procedure refers to the part of the process that a worker or a group of workers continuously complete on the same or simultaneous processing of several workpieces at a work site.

The process can be divided into hair germ manufacturing, mechanical processing, heat treatment and assembly.

The main criteria for identifying the same process are: whether the work place is changed or whether the process is continuous.

Installation

Installation refers to the part of the process that is completed by fixing and positioning the workpiece once in a machine tool or fixture.

◆ Location: Before processing, the workpiece should occupy the correct position in the machine tool or fixture. This process is called positioning.

◆ Clamping: After the workpiece is positioned, it is fixed so that the positioning position of the workpiece remains unchanged in the process of processing. The operation system is called clamping.

Workstation

Workplace refers to each position occupied by the workpiece and the movable part of the fixture or equipment together with the fixed part of the tool or equipment in order to complete a certain process content.

◆ In order to reduce the errors and time losses caused by hesitant installation, rotary worktable, rotary fixture or mobile fixture are often used in multi-station machining, so that the workpiece can be processed in several different positions in one installation.

Working steps

The process can be divided into several steps. If the machined surface remains unchanged, the cutting tool remains unchanged, the feed in the cutting parameters and the cutting speed remain basically unchanged, please step over the lower rope to complete the part of the process continuously, which is called the step.

As long as one of the three invariable factors changes, it will become a new step.

A process consists of one or more steps.

When a workpiece carries on several identical steps continuously after one clamping, it is often filled in as one step.

Using several cutters or composite cutters to process several surfaces on the same piece at the same time is called composite step. In the process documents, the compound step is regarded as a step.

Feed

The cutting tool cuts a layer of metal from the machined surface, which is called one-time feed. A step may be fed only once or several times.

production program

The output and schedule plan of products that a factory should produce during the planning period is called the production program. The quantity of qualified products manufactured by factories in one year is called annual production program and annual output. In addition to the planned quantity, the production program of a part in a product must include the spare parts rate and the average reject rate.

Production type

Production type refers to the classification of the degree of production specialization of enterprises. According to the difference of product size and production program, mechanical manufacturing production is generally divided into three types: single-piece production, finished product production and mass production.

Processing Procedure Regulations

Technological Characteristics of Various Production Types

Machining process specification is a technical document that fixes the manufacturing process and operation method of products or parts according to a certain format. Under specific production conditions, it is compiled on the basis of the most reasonable and economical principle and is used as a regulatory document to guide production after examination and approval.

Machining process rules include parts processing process flow, processing process content, cutting parameters, equipment and process equipment, man-hour quota, etc.

Machining process regulations are the most important technical documents of machinery manufacturing plants and an important part of plant rules and regulations. Their main functions are as follows:

(1) It is the basic basis for organizing and managing production.

(2) It is the main technical document guiding production.

(3) It is the raw material for new and expanded factories.

(4) It is the basic information for technological exchange and innovation.

According to the guidance technical document JB/Z338.5 of the former Ministry of Mechanical and Electronic Industry, "Process Management Guidelines and Process Planning", the types of process planning are as follows:

(1) Special Process Plans - Process Plans designed for each product and part.

(2) General process planning, which includes.

1) Typical process planning - general process planning for a group of components with similar structure.

2) Grouping process rules - the general process rules designed for each group of parts are classified into groups according to the principle of grouping technology.

3) Standard Procedures - Procedures that have been incorporated into national or factory standards.

In order to meet the needs of industrial development, strengthen scientific management and facilitate exchanges, the former Ministry of Mechanical and Electronic Industry has also formulated the Guiding Technical Document JB/Z187.3-88 Format of

Processing Procedures. According to the regulations, there are:

1) Machining Process Card: Mainly lists the whole process route, tooling equipment and working hours of parts processing, which are mostly used as production management.

2) Machining process card: the most detailed process document used to guide workers'operation. On the card, the working process sketch should be drawn, indicating the processing surface of the process and the size accuracy and roughness requirements to be achieved, the installation method of the workpiece, cutting parameters, tooling equipment, etc.

3) Process cards for standard parts or typical parts.

4) Single-axle automatic lathe adjustment card.

5) Multi-axis automatic lathe adjustment card.

6) Operational guidance card for machining process.

7) Inspection cards, etc.

The following are the assembly process regulations.

1) Process card.

2) process card.

Principles for Formulating Process Regulations

The following principles must be followed in formulating process regulations:

1) Full use of our existing production conditions must be made.

2) Parts that meet the requirements of drawings must be processed reliably to ensure the quality of products.

3) Guarantee good working conditions and improve labor productivity.

4) On the premise of guaranteeing product quality, reduce consumption and cost as much as possible.

5) Advanced technology at home and abroad should be adopted as far as possible. As the process specification is a technical document directly guiding production and operation, it should also be clear, correct, complete and unified, and the terms, symbols, codes and measurement units used must meet the relevant standards.

Major Basis for Formulating Process Regulations

In formulating the process regulations, the following raw data must be used:

1) Assembly drawings of products and working drawings of parts.

2) Production program of products.

3) The existing production conditions of the enterprise, including the production conditions or cooperative relations of the blanks, process equipment and special equipment and their manufacturing capacity, the technical level of workers and various process data and standards, etc.

4) Quality standards for product acceptance.

5) Relevant information on new technologies, new processes and development prospects of similar products at home and abroad.

Steps for formulating process regulations

The steps for formulating the machining process regulations are as follows:

1) Familiarity with and analysis of the main basis for the formulation of process regulations, and determination of the production program and types of parts.

2) Analyse the working drawings of parts and assembly drawings of products, and carry out the structural and technological analysis of parts.

3) Determine the blank, including the choice of blank type and its manufacturing method.

4) Choosing the location datum or the location base plane.

5) Drawing up the technological route.

6) Determine the equipment and process equipment needed for each process.

7) Determine process margin, process size and tolerance.

8) To determine the technical requirements and inspection methods for each main process.

9) Determine the cutting parameters and time quota of each working procedure, and make technical and economic analysis to select the best technological scheme.

10) Fill in the process documents.

10.3 工艺卡片实例与工序卡实例

减速器输出轴机械加工工艺过程卡片

机械加工工艺过程卡片		产品型号		零件图号			共页	第页
		产品名称	减速器	零件名称	减速器输出轴			
材料牌号	45	毛坯种类	棒料	毛坯外形尺寸	$\Phi60\times220$	每毛坯可制件数	1	每台件数		备注	

工序号	工序名称	工序内容	车间	工段	设备	工艺装备	工时准终	工时单件
1	备料	圆钢棒料 $\Phi60\times220$						
2	车	卡盘夹右端 (1) 粗车左端端面 (2) 钻中心孔B3.15 (3) 粗车两处 $\Phi35\pm0.008$ 外部分至 $\Phi36.7^{0}_{-0.34}\times25m$ 和 $36.7^{0}_{-0.34}\times40mm$，粗车 $\Phi52\times9$ 外圆部分至图样要求，粗车 $\Phi42\pm0.0125$ 外圆部分至中 $\Phi43.6^{0}_{-0.34}\times52mm$ 粗车 $\Phi33^{0}_{-0.34}$ 外部分外圆至 $\Phi34.4^{0}_{-0.34}\times32mm$			C6132A	45°端面车刀02R2020和90°外圆粗车刀06R1616，带护锥中心站（B型）；自定心卡盘；0.01规格的带表卡尺。		
3	车	掉头卡盘装夹左端 (1) 车右端面，保证零件总长215m (2) 钻中心孔B3.15 (3) 粗车 $\Phi28^{0}_{-0.34}$ 外图部分至 $\Phi29.5^{0}_{-0.34}\times28mm$，$M24\times1.5$-8g外圆部分至 $\Phi25.1^{0}_{-0.34}\times18mm$			C6132A	45°端面车刀02R2020和90°外圆粗车刀02R1616，带护锥中心孔(B型)；自定心卡盘；0.01规格的带表卡尺		
4	热	调质处理HBS235						
5	研	修研中心孔			C6132A	60°锥磨头		
6	车	两顶尖装夹工作 (1) 半精车两处 $\Phi35\pm0.008$ 外圆部分至 $\Phi35^{0}_{-0.34}\times25mm$ 和 $35.3^{0}_{-0.34}\times40mm$，半精车 $\Phi42\pm0.0125$ 外圆部分至 $\Phi42.2^{0}_{-0.34}\times52mm$，半精车 $\Phi33^{0}_{-0.34}$ 外圆部分至 $\Phi33^{0}_{-0.34}\times32mm$ (2) 车R2圆弧，$\Phi42\times4$ 至图样尺寸 (3) 车两处 $3\times0.05mm$ 的越程槽 (4) 倒角C1			C6132A	01R161670外圆车刀；04R2012切槽刀；02R2020 45°端面车刀；0.01规格带表卡尺；两顶尖，鸡心夹头等		

机械加工、成型和安装工艺

续表

		机械加工工艺过程卡片	产品型号		零件图号		共 页	第 页
			产品名称	减速器	零件名称	减速器输出轴		
7	车	掉头两顶尖装夹工件 (1) 半精车$\Phi28^{0}_{-0.034}$外圆部分至$\Phi28^{0}_{-0.034}\times28mm$，半精车M24×1.5-8g外圆部分至$\Phi28^{0}_{-0.34}\times18mm$ (2) 车一处3mm×1mm的退刀槽 (3) 倒角C1、C1.5			C6132A		01R161670外圆精车到刀，04R2012切槽刀，02R202045°端面车刀；0.01规格带表卡尺；两顶尖，鸡心夹头等	
8	车	两顶尖装夹工件 (1) 精车$\Phi42\pm0.0125$外圆部分和$\Phi28^{0}_{-0.34}$外圆部分至图样尺寸 (2) 车M24×1.5-8g螺纹至图样尺寸			C6132A		01R161670外圆精车刀，螺纹车刀；两顶尖，鸡心夹头等；0.01规格代表卡尺，螺纹环规	
9	铣	铣床附件装夹工件 (1) 铣$\Phi42\pm0.0125$外圆面上的键槽至图样尺寸 (2) 铣$\Phi28^{0}_{-0.34}$外圆面上的键槽至图样尺寸			X6025		不同直径直柄麻花钻两把，不同宽度直柄键槽铣刀两把；V形块抱钳；0.02规格游标卡尺，指示表	
10	磨	两顶尖装夹工件 (1) 粗磨两处$\Phi35\pm0.008$外圆及轴肩至图样尺寸要求 (2) 精磨两处$\Phi35\pm0.008$外圆及轴肩至图样尺寸要求			MW1432		单面凹带锥砂轮；顶尖，鸡心夹头；外径千分尺，表面粗糙度标准样块	
11	检	检验						

减速器输出轴零件机械加工工序卡片

机械加工工序卡片	产品型号		零件图号				
	产品名称	减速器	零件名称	减速器输出轴		共 页	第 页
			车间	工序号	工序名称	材料牌号	
				7	车	45	
			毛坯种类	毛坯外形尺寸	每毛坯可制件数	每台合计	
			设备名称	设备型号	设备编号	同时加工件数	
			卧式车床	C8132A		1	
			夹具编号	夹具名称		切削液	
				前后顶尖，鸡心夹头			
				工位器具编号	工位器具名称	工序工时	
						准终	辅助

工步号	工步内容	工艺装备	主轴转速(r/min)	切削速度(m/min)	进给量(mm/r)	背吃刀量(mm)	进给次数	工步工时	
								机动	辅助
1	半精车 $\Phi 28^{0}_{-0.34}$ 外圆部分至 $\Phi 28^{0}_{-0.34} \times 28mm$，半精车 M24×1.5-8g 外圆部分至 $\Phi 23.8^{0}_{-0.34} \times 18mm$		460	50	0.2	0.4			
2	车一处3mm×1mm的退刀槽		460	50	0.1		1		
3	倒角C1和C1.5		460	50	0.1		1		

							设计日期	审核日期	标准化日期	合同日期
标记	处数	更改文字编号	签字	日期	标记	处数	更改文字编号	签字	日期	

减速器输出轴零件机械加工工序卡片

机械加工工序卡片	产品型号		零件图号			
	产品名称	减速器	零件名称	减速器输出轴	共 页	第 页

	车间	工序号	工序名称	材料牌号
		10	磨	45
	毛坯种类	毛坯外形尺寸	每毛坯可制件数	每台合计
	设备名称	设备型号	设备编号	同时加工件数
	万能外圆磨床	MV1432		1
	夹具编号	夹具名称		切削液
		前后顶尖，鸡心夹头		
	工位器具编号		工位器具名称	工序工时 准终 / 辅助

工步号	工步内容	工艺装备	主轴转速(r/min)	切削速度(m/min)	进给量(mm/r)	背吃刀量(mm)	进给次数	工步工时 机动 / 辅助
1	粗磨两处Φ35±0.008外圆及轴肩直径Φ35±0.058		213	23	25.2	0.009		
2	精磨两处Φ35±0.008外圆及轴肩至图样尺寸要求		324	35	16.8	0.004		
			设计日期	审核日期	标准化日期		合同日期	
标记	处数	更改文字编号	签字	日期	标记	处数	更改文字编号	签字	日期

第十一章 装配工艺基础

11.1 装配的概念

机械产品一般是由许多零件和部件组成的，根据规定的技术要求，将若干个零件结合成部件或将若干个零件和部件结合成产品的过程，称为装配；前者称为部件装配，后者称为总装配。

为了制造出合格的产品，必须抓住三个主要的环节：第一，产品结构设计的正确性，它是保证产品质量的先决条件；第二，组成产品的各零件的加工质量，它是产品质量的基础；第三，装配质量和装配精度，它是产品质量的保证。

装配过程并不是将合格零件简单连接起来的过程，而是根据各级部件装配和总装配的技术要求，通过校正、调整、平衡、配作以及反复试验来保证产品质量的复杂过程。若装配不当，即使零件质量都合格，也不一定制造出合格的产品；反之，即使零件质量不太好，只要在装配中采取了合适的工艺措施，也能使产品达到或基本达到规定的质量要求。

11.2 装配工作的基本内容

清 洗

为了去除零件表面或部件中的油污及机械杂质，保证产品的质量和延长使用寿命，装配前要进行零、部件的清洗。常用的清洗方法有擦洗、浸洗、喷洗和超声波清洗等，常用的清洗液有煤油、汽油、碱液及各种化学清洗液等。

连 接

将两个或两个以上的零件结合在一起称为连接，连接的方式一般有两种：可拆卸连接和不可拆卸连接。

可拆卸连接是指相互连接的零件拆卸时不受任何损坏，而且拆卸后能够重新装配。常见的有螺纹连接、键连接、销钉连接等。

不可拆卸连接是指相互连接的零件在连接后不可拆卸，若要拆卸则会损坏某些零件。常见的有焊接、铆接和过盈连接等。过盈连接大多应用于轴、孔的配合，可使用压入法、热胀法和冷缩配合法实现。

校正、调整与配作

在产品装配过程中，特别是在单件小批生产条件下，为了保证装配精度，常需进行一些校正、调整和配作工作，这是因为完全靠零件精度来保证装配精度往往是不经济的，有时甚至是不可能的。

机械加工、成型和安装工艺

校正是指产品中相关零、部件相互位置的找正、找平，并通过各种调整方法达到装配精度。

调整是指调节相关零部件的相互位置，除配合校正所作的调整外，还有各运动副间隙，如轴承间隙、导轨间隙、齿轮齿条间隙的调整等。

配作是指配钻、配铰、配刮和配磨等在装配过程中所附加的一些钳工和机械加工工作，如连接两零件的销孔，就必须待两零件的相互位置找正后再一起钻销钉孔，然后打入定位销钉，这样才能确保其相互位置准确。

配刮是指将配合面涂上红丹粉，然后使运动副作相对运动，根据配合表面的接触情况将高点刮去，如此反复，直至达到要求。配刮可提高两结合面之间的接触精度和运动副的运动精度，有利于润滑油的储存，可提高零件的耐磨性。因此在机器的装配和修理中常用到配刮，但其生产效率低，劳动强度大，应尽量"以刨代刮、以磨代刮"。

必须指出，配作是在校正、调整的基础上进行的，只有经过认真的校正、调整后才能进行配作。校正、调整与配作虽有利于保证装配精度，但会影响生产效率，不利于流水装配作业。

平 衡

对于转速高、运转平稳性要求较高的机器，为了防止使用中出现振动，装配时必须对有关旋转的零件进行平衡，必要时还要对整机进行平衡。

平衡有静平衡和动平衡两种方法。对于直径较大、长度较小的零件（如带轮和飞轮），一般只需进行静平衡；对于长度较大的零件（如电动机转子和机床主轴），则需进行动平衡。

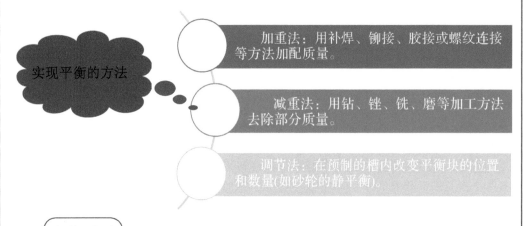

实现平衡的方法

加重法：用补焊、铆接、胶接或螺纹连接等方法加配质量。

减重法：用钻、锉、铣、磨等加工方法去除部分质量。

调节法：在预制的槽内改变平衡块的位置和数量(如砂轮的静平衡)。

验收试验

机械产品装配完成后应根据其质量验收标准进行全面的验收试验，各项验收指标合格后才可涂装、包装、出厂。机械产品的种类不同，其验收技术标准也不同，验收试验的方法也就不同。

11.3 装配工艺及方法

装配工艺是规定产品及部件的装配顺序、装配方法、装配技术要求和检验方法及装配所需的设备、工夹具、时间、定额等的技术文件。常见的装配方法有互换装配法、分组装配法、修配装配法和调整装配法四种。

互换装配法

互换装配法是指在装配过程中，各零件不需挑选、修配和调整即可达到装配精度要求的一种方法。互换装配法的实质是用控制零件的加工误差来保证产品的装配精度。根据互换程度的不同，互换装配法可分为完全互换法和不完全互换法两种。

分组装配法

分组装配法也称为分组互换法、选配法。这种方法就是当装配精度要求极高、零件制造公差限制很严，致使零件几乎无法加工时，可将零件的公差放大到经济可行的程度，然后按实测尺寸将零件分组，再按对应组分别进行装配，以达到装配精度要求的一种装配方法。

修配装配法

修配装配法是将装配尺寸链中的各组成环按经济精度加工，装配时，通过改变尺寸链中某一预定的组成环（修配环）尺寸的方法来保证装配精度。由于对修配环的修配是为了补偿其他各组成环的累积误差，故又称为补偿环。这种方法的关键是确定修配环在加工中的实际尺寸，使修配环有足够的、而且是最小的修配量。

修配装配法适用于成批生产中封闭环公差要求较严、组成环多或单件小批生产中封闭环公差要求较严、组成环较少的场合。

采用修配装配法时，装配尺寸链一般用极值法计算。

调整装配法

调整装配法的实质与修配装配法相同，也是将尺寸链中各组成环的公差值放大，使其按经济精度制造。装配时，预先选定尺寸链中的某一组成环作为调整环，采用调整的方法改变其实际尺寸或位置，使封闭环达到规定的公差要求。预先选定的环成为"调整环"，是用来补偿各组成环因公差放大而产生的累计误差。常见的调整方法有可动调整法、固定调整法和误差抵消调整法三种。

11.4 装配精度

为了使机器具有正常的工作性能，必须保证其装配精度。机器的装配精度通常包含以下三个方面的含义：

(1) 相互位置精度：指产品中相关零部件之间的距离精度和相互位置精度，如平行度、垂直度和同轴度等。

(2) 相对运动精度：指产品中有相对运动的零部件之间在运动方向和相对运动速度上的精度，如传动精度、回转精度等。

(3) 相互配合精度：指配合表面间的配合质量和接触质量。

装配精度的确定原则

(1) 对于一些标准化、通用化和系列化的产品，如通用机床和减速器等，它们的装配精度可根据国家标准、部颁标准或行业标准来确定。

(2) 对于没有标准可循的产品，可根据用户的使用要求，参照经过试验的类似产品或部件的已有数据，采用类比法确定。

(3) 对于一些重要产品，需要经过分析计算和试验研究后才能确定。

装配精度与零件精度的关系

机械产品是由许多零件组成的，零件的精度特别是关键零件的精度对整机的装配精度有直接的影响。要保证整机的装配精度，就必须控制相关零件的加工精度。一般来说，装配精度要求越高，与此项装配精度有关的零件的加工精度要求也就越高。因此，产品的装配精度与零件的加工精度密切相关。零件的加工精度是保证装配精度的基础，但装配精度并不完全取决于零件的加工精度。装配精度的合理保证，应从产品结构、机械加工和装配工艺等方面综合考虑。装配尺寸链的分析是进行综合考虑的有效手段。

11.5 装配的组织形式

根据产品结构特点和生产批量的大小，装配工作可以采用不同的组织形式，一般有固定式和移动式两种。

1. 固定式装配

固定式装配是将产品或部件的全部装配工作安排在一个固定的工作地点进行。在装配过程中产品的位置不变，装配所需的零件也集中放在工作地点附近。根据产品的结构和生产类型，固定式装配有三种形式。

(1) 集中固定式装配。全部装配工作由一组工人在一个工作地点集中完成，属于集中固定式装配。这种装配组织形式要求工人技术水平高，而且装配时间长，多适用于单件小批生产。

(2) 分散固定式装配（又称多组固定式装配）。这种装配形式是把产品的全部装配过程分解为组部件装配和总装配,分别在多个工作地点进行。各部件的装配和产品的总装由几组工人在不同的工作地点分别进行。这种组织形式可使装配操作专业化,装配周期短,生产场地的使用率和生产效率较高。

(3) 产品固定式流水装配。这种装配形式是将装配过程分成若干个独立的装配工序,分别由几组工人负责。各组工人按工序顺序依次到各装配地点对固定不动的装配对象进行本组所要求的进行装配。这是固定装配的高级形式,工人专业化程度高,产品质量稳定,装配周期短,适用于笨重产品的成批生产。

2. 移动式装配

移动式装配是装配工人和工作地点固定不变,装配对象连续地通过每个工作地点,在一个工作地点完成一个或几个工序,在最后一个工作地点完成装配工作。这种装配方式的特点是各装配时间重合或部分重合,因而装配周期短,工人专业化程度高,工作地点固定,降低了劳动强度。

11.6 装配工艺规程

制定装配工艺过程的基本原则

(1) 保证产品的装配质量,以延长产品的使用寿命;

(2) 合理安排装配顺序和工序,尽量减少钳工手工劳动量,缩短装配周期,提高装配效率;

(3) 尽量减少装配占地面积;

(4) 尽量减少装配工作的成本。

制定装配工艺规程的步骤

1. 研究产品的装配图及验收技术条件

(1) 审核产品图样的完整性、正确性;

(2) 分析产品的结构工艺性;

(3) 审核产品装配的技术要求和验收标准;

(4) 分析和计算产品装配尺寸链。

2. 确定装配方法与组织形式

(1) 装配方法的确定:主要取决于产品结构的尺寸大小和重量,以及产品的生产纲领。

(2) 装配组织形式:

① 固定式装配:全部装配工作在一固定的地点完成。适用于单件小批生产和体积、重量大的设备的装配。

② 移动式装配:将零部件按装配顺序从一个装配地点移动到下一个装配地点,分别完成一部分装配工作,各装配点工作的总和就是整个产品的全部装配工作。适用于大批量生产。

3. 划分装配单元,确定装配顺序

(1) 将产品划分为套件、组件和部件等装配单元,进行分级装配;

(2) 确定装配单元的基准零件;

(3) 根据基准零件确定装配单元的装配顺序。

4. 划分装配工序

(1) 划分装配工序,确定工序内容(如清洗、刮削、平衡、过盈连接、螺纹连接、校正、检验、试运转、油漆、包装等);

(2) 确定各工序所需的设备和工具;

(3) 制定各工序装配操作规范,如过盈配合的压入力等;

(4) 制定各工序装配质量要求与检验方法;

(5) 确定各工序的时间定额,平衡各工序的工作节拍。

5. 编制装配工艺文件

常用的装配工艺过程卡片格式如表11-1所示。根据对装配内容的分析编制装配工艺过程卡片,用于指导生产

表11-1 装配工艺过程卡片

装配工艺过程卡片		产品型号		零部件图号		共( )页
		产品名称		零部件名称		第( )页
工序号	工序名称	工序内容	装配部	工艺装备	辅助材料	时间定额/min
标记	处数	文件号	签字	设计	审核	会签

第十二章 车间布局与 5S 管理

12.1 概述

在之前的学习中，我们已经了解到计划生产零件的年产量叫作生产纲领，生产纲领对一个生产型企业的方方面面都有着巨大的影响，而车间布局就是受生产纲领影响非常大的一个方面。和车间布局相关的一个重要企业管理手段"5S 现场管理"也会在本章进行简要介绍。

12.2 车间布局

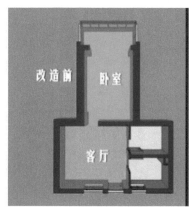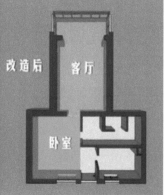

众所周知，同样的一套房屋，不合理的布局会显得空间狭小，不易于储物，甚至造成居住人心情低落；而合理的布局，会让有限的空间发挥出更大的利用效率，物品摆放合理，住在里面的人才会舒服。

车间布局的优劣也会极大地影响生产效率和员工士气。

对于一个企业来说，无论厂房、车间或租或购，厂房车间面积都是生产成本的重要组成部分。每一平方米的厂房都应该最少体现其一平方米的价值，如果车间布局合理，每一平方米会产生更大价值，但如果布局不合理，不仅会浪费车间面积，还会陷入生产效率低下的困境。

机械加工、成型和安装工艺

如何合理的为车间进行布局呢？通常情况下要遵循的原则有哪些呢？

合理的布局生产方式是对传统企业大批量生产的变革，颠覆了企业功能式布局，消除经验式的管理桎梏。流动化生产的模式设计从流动化管理开始，布局结构的设计应符合流动的要求。因此需要企业管理团队对生产车间的设备、工装、人员、物料、质量与技术等因素进行全方位的分析和研究，探讨出最适合当前状态的流动方案，系统的流动化布局规划应该关注以下原则。

统筹规划原则

- 人员的技能和多能化。
- 设备及工装的稳定及快速切换能力。
- 物料的准时配送。
- 工艺及质量标准的过程执行。
- 现场5S管理的日常化。

> 七大原则慢慢说来！

标准化原则

标准化作业能维持员工作业的稳定和高效，避免制造过多步行距离远、手动作业的浪费。

标准化作业是流动化管理及正常生产运行的必要条件。

标准化作业的目的

- 作业顺序一致化。
- 逆时针方向操作。
- 进行适当的作业组合。
- 明确作业循环时间。
- 明确在制品数量。

信息流畅原则

信息是流动化管理运行的指令中心,是企业制造过程流畅运行的前提。信息流畅说明生产计划指令明确,而且便于信息的横向和纵向传递,因此制造过程就比较稳定。

物流流畅原则

生产组织应能够实现物流流畅,从而保证各工序的生产和物料供应保持同步,促进各工序生产节拍的协调一致,有利于实现"在必要的时刻得到必要的零件"。

用户导向原则

实现合理的布局生产方式就需要各道加工工序"零缺陷、零故障",并且一次把事情做对。如果制造过程中出现缺陷或不稳定的因素,要么停掉生产线,组织资源进行改善;要么强行把有缺陷的在制品废弃,无论何种选择都将引起成本的上升。因此精益化的布局生产方式要求每一道工序严格控制工作质量,做到质量在过程中控制,遵循用户导向原则。

消除浪费原则

合理的布局生产方式的目的是减少在制品,使生产过程中存在的浪费现象暴露出来并不断排除,使成本下降。

安全为重原则

流动化管理的过程中要确保员工的作业安全。由于布局的紧凑,导致员工的工作空间和作业场所发生变化,应站在员工作业的立场上考虑安全隐患;同时对员工具体的作业过程进行分析,消除影响员工安全的隐患因素。需要考虑的要素如下。

◆加工点远离双手可达区域　　◆防止误启动
◆作业时容易步行　　　　　　◆蒸汽、油污、粉屑防护
◆去除突出物　　　　　　　　◆现场照明、换气、温度、湿度

How to arrange the workshop reasonably? What principles do we usually follow?

Reasonable layout of production mode is the change of mass production of traditional enterprises, which subverts the functional layout of enterprises and eliminates the shackles of empirical management. The mode design of the flow production starts with the flow management, and the layout structure design should meet the flow requirements. Therefore, it is necessary for enterprise management team to conduct a comprehensive analysis and Research on the factors such as equipment, tooling, personnel, materials, quality and technology in the production workshop, and to explore the most suitable flow plan for the current situation. Systematic flow layout planning should pay attention to the following elements:

1. Overall planning principles
- Skills and versatility of personnel
- Stability and fast switching capability of equipment and equipment
- Material Delivery on Time
- Process execution of process and quality standards
- Daily management of 5S on site

2. Standardized Operational Principles

Standardized work can maintain the stability and efficiency of employees' work, avoid the waste of excessive production, long walking distance and manual work. Standardized operation is a necessary condition for mobile management and normal production operation. Its purpose is as follows:

1. Work order consistency
2. Counterclockwise operation
3. Proper combination of operations
4. Make clear the working cycle time
5. Define the quantity of products in process

3. The principle of information fluency

Information flow is the command center of mobile management and the prerequisite of smooth operation of enterprise manufacturing process. Smooth information indicates that the production planning instructions are clear, and it is convenient for the horizontal and vertical transmission of information, so the manufacturing process is relatively stable. The effects of production information fluency on enterprises are as follows:
- Realizing pull-type production
- Visualization of Production Performance
- Facilitate batch processing

4. Logistics Fluency Principle

Production organization should be able to achieve smooth logistics, so as to ensure the synchronization of production and material supply in each process, promote the coordination of production rhythm in each process, and help to achieve the "necessary parts at the necessary time".

5. user orientation

To achieve a reasonable layout of production methods, it is necessary for each processing process to "zero defects, zero failures" and make it right at one time. If there are defects or unstable factors in the manufacturing process, either stop the production line, organize resources to improve, or forcibly abandon the defective in-process products, regardless of the choice will lead to increased costs. Therefore, the lean layout production mode requires that each process strictly control the quality of work, to achieve quality control in the process, and to follow the user-oriented principle. The meaning of user orientation is:

◆ Each process is the user of the previous process.
◆ Each process is the supplier of the subsequent process.
◆ Each process only accepts the qualified products of the previous process.
◆ Only qualified products are produced in each process.
◆ Each process only provides qualified products to the subsequent process.

The specific method to follow the principle of internal users is to carry out self-inspection and mutual inspection, and to produce in strict accordance with the process operation specifications.

6. Principle of Eliminating Waste

The purpose of rational layout of production mode is to reduce in-process products, expose waste phenomena in production process and constantly eliminate them, so as to reduce costs. These waste phenomena include:

◆ WIP surplus
◆ supply dragging
◆ Long time to troubleshoot equipment
◆ Impaired Information Exchange
◆ Poor process discipline

By improving the flow layout, we can achieve a lean state of the flow and eliminate the waste of the process.

7. Safety as the Principle

In the process of mobile management, we should ensure the safety of employees' work. Because of the compact layout, the workspace and workplaces of employees will change. We should consider the potential safety hazards from the standpoint of employees' work. At the same time, the specific work process of employees is analyzed to eliminate the hidden danger factors affecting the safety of employees. The following factors need to be considered:

- ◆ The processing point is far from the reachable area of hands.
- ◆ Easy to walk while doing homework
- ◆ Remove protrusions.
- ◆ Prevent misoperation.
- ◆ Protection of Steam, Oil and Powder
- ◆ Field lighting, ventilation, temperature and humidity

12.3　5S 管理

人人随处乱扔垃圾，没有人来捡。
（三流企业）

由专人负责捡垃圾。
（二流企业）

每个人都不会乱扔垃圾，维护环境清洁。
（一流企业）

我们为什么要学习5S现场管理法呢？

5S 现场管理法是现代生产组织系统中十分重要的管理方法。

现代生产组织系统推广的目的是打造"精益工厂""精益生产"，从而达到效率最大化、消耗最小化、利润最大化。现代生产组织系统还包括：柔性生产系统、TQM 精益质量保证、生产与物流规划、TPM 全面设备维护、产品开发设计系统、现代 IE 运用、均衡化同步化等。而 5S 现场管理法是上述所有管理方法运行的基石，没有有效运用 5S 现场管理法之前，其他管理方法都只是空谈。

5S管理从何而来？

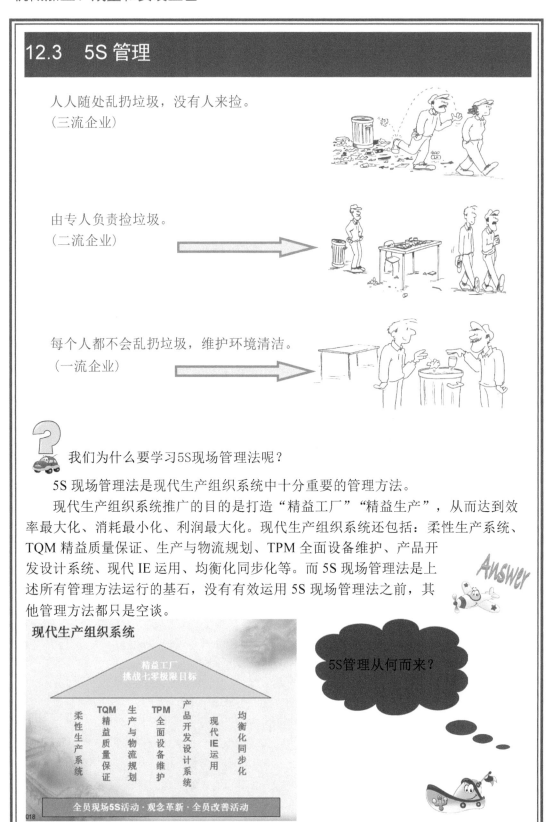

第四篇　机械加工工艺规程与车间管理

 5S管理的起源

"5S"起源于日本，是指在生产现场对人员、机器、材料、方法、环境等生产要素进行有效管理，这是日本企业独特的一种管理办法。

1955年，日本的5S的宣传口号为"安全始于整理，终于整理整顿"。当时只推行了2S，其目的仅为了确保作业空间和安全。后来因生产和品质控制的需要而又逐步提出了3S，也就是清扫、清洁、修养，从而使应用空间及适用范围进一步拓展。到了1986年，日本的5S的著作逐渐问世，从而对整个现场管理模式起到了冲击的作用，并由此掀起了5S的热潮。

日本企业将5S运动作为管理工作的基础，推行各种品质的管理方法，第二次世界大战后，产品品质得以迅速地提升，奠定了经济大国的地位。而在丰田公司的倡导推行下，5S对于塑造企业的形象、降低成本、准时交货、安全生产、高度的标准化、创造令人心旷神怡的工作场所、现场改善等方面发挥了巨大作用，逐渐被各国的管理界所认识。随着世界经济的发展，5S已经成为工厂管理的一股新潮流。

5S即日文的：整理（Seiri）、整顿（Seiton）、清扫（Seiso）、清洁（Seiketsu）、素养（Shitsuke），这五个日语单词的罗马拼音均以"S"开头，英语也是以"S"开头，所以简称5S。

这五个单词，又被称为"五常法则"或"五常法"，现在已经最多被扩展为"12S"，这是在5S的基础上增加了：安全（Safety），即形成了"6S"；节约（Save），形成了"7S"；习惯化（Shiukanka），这就是"8S"；服务（Service）；形成"9S"；坚持（Shikoku），形成了"10S"。

有的企业甚至推行"12S"，但是万变不离其宗，都是从"5S"衍生出来的。其实，无论怎么变化，5S其实是一种管理思想和文化。

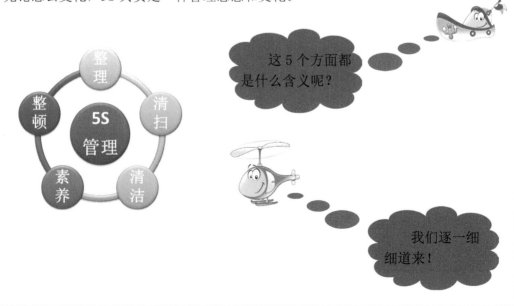

机械加工、成型和安装工艺

下文中会遇到的名词：
现场：指从事生产、工作、试验的所有场所，包括车间、仓库、办公室等直接场所也包括卫生间、杂物间以及通道和花园等"间接场所"。
物品：泛指出现在现场的一切人、事、物。

整理（Seiri）

◆定义：区分要与不要的物品，现场只保留必要的物品。

◆目的：

(1) 腾出空间，改善和增加作业面积；

(2) 减少磕碰的机会，保障安全，提高质量；

(3) 消除管理上的混放、混料等差错事故，防止误用、误送；

(4) 现场无杂物，行道通畅，提高工作效率，塑造清爽的工作场所；

(5) 改变作风，提高工作情绪。

◆关键点：

扔掉物品与保存物品同样重要，要有决心断然处置不必要的物品。把要与不要的人、事、物分开，再将不需要的人、事、物加以处理，对生产现场的现实摆放和停滞的各种物品进行分类，区分什么是现场需要的，什么是现场不需要的；其次，对于车间里各个工位或设备的前后、通道左右、厂房上下、工具箱内外，以及车间的各个死角，都要彻底搜寻和清理，达到现场无不用之物。

◆难点：如何区分要与不要的物品。

整顿（Seiton）

◆定义：必需品定点、定位、定量地摆放整齐有序，并标示清楚。

◆目的：

(1) 消除寻找物品的时间（要用的东西随即可取得）；

(2) 整整齐齐的工作环境，保障生产安全；

(3) 工作场所一目了然。

◆关键点：

把需要的人、事、物加以定量、定位。通过前一步整理后，对生产现场需要留下的物品进行科学合理的布置和摆放，以便用最快的速度取得所需之物，在最有效的规章、制度和最简洁的流程下完成作业，这是提高效率的基础。

◆难点：

摆放有序，标示清晰，人人一目了然。① 物品摆放要有固定的地点和区域，以便于寻找，消除因混放而造成的差错；② 物品摆放目视化，使定量装载的物品做到过目知数；③ 摆放不同物品的区域采用不同的色彩和标记加以区别。

第四篇　机械加工工艺规程与车间管理

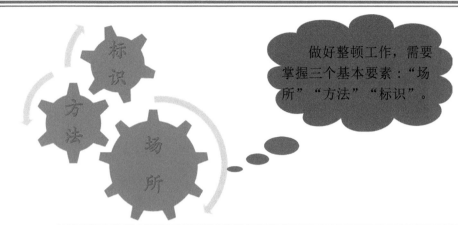

整顿三要素之"场所":指的是物品放置的场所。
(1) 物品的放置场所原则上要100%设定。
(2) 物品的保管要定位、定容、定量。
(3) 生产线附近只能放置真正需要的物品(参照下面场所确定表)。

物品使用情况及处理方式

	使用频率	处理方法	建议场所
不用	全年一次也未用	废弃或特别处理	待处理区
少用	平均2～12月用1次	分类管理	工具室、仓库
普通	1～2个月使用1次或以上	置于车间内	各摆放区
常用	1周使用数次 1日使用数次 每小时都使用	工作区内、随手可得	如机台旁、流水线旁

总之,所谓"场所"强调的是物品摆放地点要科学合理。根据物品使用的频率,经常使用的东西应放得近些(如放在作业区内),偶尔使用或不常使用的东西则应放得远些(如库房中)。

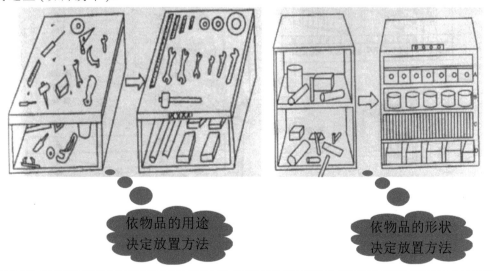

203

整顿三要素之"方法":指的是物品放置的方法。
(1) 易取。
(2) 不超出所规定的范围。
(3) 在放置方法上多下工夫。

针对不同物品所采用的"方法"应该因地制宜、对症下药、随机应变,不能墨守陈规的一种"方法"应用所有物品。

整顿三要素之"标识":指的是物品的标识。
(1) 放置场所和物品原则上一对一标识。
(2) 现物的标识和放置场所的标识。
(3) 某些标识方法全公司要统一。

清楚、明确的标识一目了然

一对一标识

清晰明确的标识,是提高效率的有效手段,也可以让新员工更快地接受工作环境和投入工作中。

定位 → 定容 → 定量

- 把物品放置在确定的位置。
- 把物品用规定的容器进行放置。
- 把物品放置合适的数量。

第四篇 机械加工工艺规程与车间管理

整顿三定原则：
　　整顿的三定原则是定位、定容、定量，是为了使人可以一眼看清什么东西在哪里，有多少，处于什么状态，以确立用眼观察（可视）的管理活动。

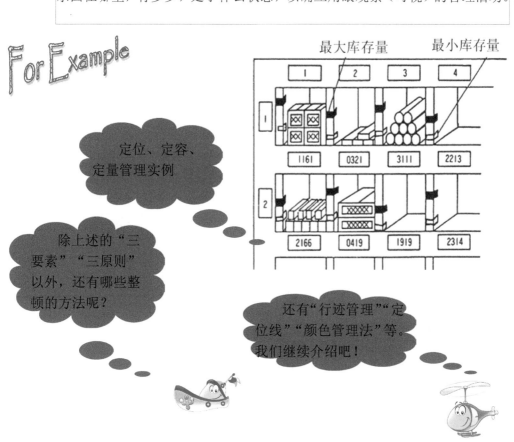

定位、定容、定量管理实例

除上述的"三要素""三原则"以外，还有哪些整顿的方法呢？

还有"行迹管理""定位线""颜色管理法"等。我们继续介绍吧！

　　行迹管理：就是根据物品或工具的"形"来管理归位的一种方法。它是根据物品的形状进行归位，能够做到对号入座，能够达到一目了然、方便取放的效果。例如下图中红色图形就是归位依据。

机械加工、成型和安装工艺

定位线：是用于地面物品的定位，视实际情况可以采用实线、虚线或四角定位线等形式，一般线宽 3～6 cm。定位线通常采用黄色线条；某些物品为了特别区分（如清洁工具、垃圾箱、凳椅等），可使用白色。

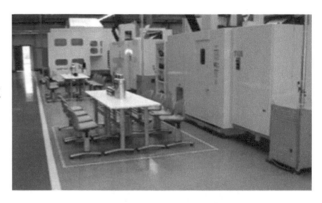

对消防器材或危险物品的定位（如乙炔气瓶），为达到警示效果，应使用红色线条，前方禁止摆放的区域（如消防栓前、配电柜前）应使用红色斑马线。

对于形状规则的小物品定位时，可采用四角定位法，其中物品角和定位角线间距应在 2～4 cm。

类别	区域线			标识牌	字体
	颜色	宽带	线型		
待检区	蓝色	50 mm	实线	蓝色	白色，黑体
待判区	白色	50 mm	实线	白色	黑色，黑体
合格区	绿色	50 mm	实线	绿色	白色，黑体
不合格区、返修区	黄色	50 mm	实线	黄色	白色，黑体
废品区	红色	50 mm	实线	红色	白色，黑体
毛坯区、展示区、培训区	黄色	50 mm	实线		
工位器具定置点	黄色	50 mm	实线		
物品临时存放区	黄色	50 mm	实线		"临时存放"字样

第四篇 机械加工工艺规程与车间管理

整顿是"5S 管理"中非常重要的一个环节，它的实施方法多种多样，要根据现场物品的具体类型来进行灵活选择，做到活学活用才能做到合理有效的进行现场整顿。

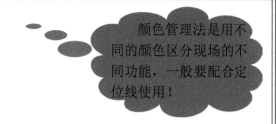

颜色管理法是用不同的颜色区分现场的不同功能，一般要配合定位线使用！

清扫（Seiso）

◆定义：清除作业区域的垃圾和脏污。

◆目的：

(1) 保持令人心情愉快、干净亮丽的环境；

(2) 减少脏污对品质的影响；

(3) 减少工业伤害事故。

◆关键点：

将现场的污垢去除，使异常的发生源很容易发现，是实施自主保养的第一步，这是为高效率完成工作、稳定品质提供条件。

◆难点：

(1) 自己使用的物品，如设备、工具等，要自己清扫，而不要依赖他人；领导者带头来做，不增加专门的清扫工，划分责任区，明角、暗角都要清扫。

(2) 要用心来做，对设备的清扫着眼于对设备的维护保养。清扫设备要同设备的点检结合起来，清扫即点检；清扫设备要同时做设备的润滑工作，清扫也是保养，并且规定例行清扫时间。

(3) 清扫也是为了改善。当清扫地面发现有飞屑和油水泄漏时，要查明原因，并采取措施加以改进。

◆清扫还需要注意以下事项：

清扫不是只在规定时间进行突击清扫，而是平时见到污渍和脏物也要马上进行处置；对于清扫对象过高、过远、手不容易够着的死角也要使用相应工具进行彻底清扫，而绝不能很少或干脆就不清扫； 当清扫工具太简单，造成许多脏物无法清除时，要及时更换或购置适合的清扫工具，清扫工具应该尽量不摆放在显眼的地方有碍美观，应集中放置于现场外围或工具柜内。

清洁(Seiketsu)

◆定义：维护以上"3S"的实施成果，保持整洁、干净、美化的工作环境，使之制度化，规范化，习惯化。

◆目的：

(1) 维持和巩固整理、整顿、清扫活动获得的成果；

(2) 保持工作现场清洁有序的状态。

◆关键点：

要清楚前"3S"是动作，清洁是结果。通过对前"3S"的坚持与深入，从而消除发生安全事故的根源，创造一个良好的工作环境，使职工能愉快地工作。

◆难点：

(1) 车间环境不仅要整齐，而且要做到清洁卫生，保证工人身体健康，提高工人劳动热情；

(2) 不仅物品要清洁，而且工人本身也要做到清洁，如工作服要清洁，仪表要整洁，及时理发、刮须、修指甲、洗澡等；

(3) 工人不仅要做到形体上的清洁，而且要做到精神上的"清洁"，待人要讲礼貌，要尊重别人；

(4) 要使环境不受污染，进一步消除浑浊的空气、粉尘、噪音和污染源，消灭职业病；

(5) 领导者经常带头巡视，查结果、追源头，上级关心下级才有责任心。

◆对于清洁的理解中经常出现下列两个错误：

(1) 出于小团体的荣誉，为了应付检查评比经常搞突击性卫生打扫。

(2) 简单地停留在扫干净的认识上，以为只要扫干净就是清洁了，结果除了干净之外，其他方面并没有多大的改善。

素养(Shitsuke)

◆定义：人人依规定行事，按章操作，养成良好习惯。每个人都成为有教养的人。

◆目的：

(1) 培养具有良好习惯、遵守规则的员工。

(2) 营造团队精神。提升"人的品质"，培养对任何工作都讲究、认真的人。

◆关键点：

5S活动始于素养，终于素养。素养的实践始自内心而行之于外，素养比纪律要求更高。努力提高员工的自身修养，使员工养成良好的工作、生活习惯和作风，让员工能通过实践5S获得人身境界的提升，与企业共同进步，是5S活动的核心。

◆难点：

素养的形成是一个长期的过程，需要不断地宣导再宣导，要求再要求。推行晨会制度，对员工进行在职教育、制度宣讲与考核以及全员做工前操、制定礼仪守则都是素养推行的实施方法。

第四篇　机械加工工艺规程与车间管理

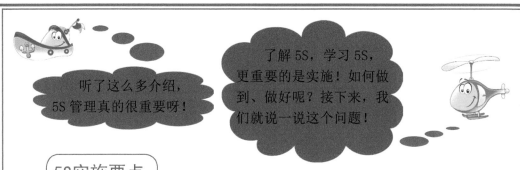

5S实施要点

整理	正确的价值意识——"使用价值"，而不是"原购买价值"。
整顿	正确的方法——"3要素""3定"+整顿的方法。
清扫	责任化——明确每个工作岗位的5S责任。
清洁	制度化及考核——稽查、竞争、奖罚。
素养	长期化——5S时间；晨会、礼仪守则。

5S实施工具

　　5S活动在进行过程中会运用大量实施工具对实施效果进行明确、检查和对比，以改善效果。常用的实施工具有红牌作战、目视管理、检查表和定点拍摄。

◆红牌作战：红牌作战是指在5S现场找到问题并悬挂红牌，让大家都明白并积极去改善，从而达到整理整顿的目的。	责任部门		希望完成日	
	问题描述：			
	对策：			
	完成日		责任人		审核	
	验收结果：			
	验收日		验收人		审核	

★红牌作战注意事项：

（1）不是处罚形式。

（2）频率不宜过高，一般一月一次，给予一定的整改时间。

（3）对于挂上红牌的物品，要说明具体原因和明确的处理方法，如重新检验入库／改作他用／降级使用／报废／变卖等，便于整改。

（4）要规定明确时间，限期整改。

（5）对象可以是材料、产品、机器、设备、空间、办公室、文件、档案等，但绝对不能针对人。

机械加工、成型和安装工艺

◆目视管理：是利用形象直观而又色彩适宜的各种视觉感知信息来组织现场生产活动，达到提高劳动生产率的一种管理手段，也是一种利用视觉来进行管理的科学方法。

★目视管理的优点：无论是谁，都能判断是好是坏（异常）；能迅速判断，精度高；判断结果不会因人而异。

图中多处定位线、颜色管理都属于目视管理。

对于设备上的一些状态线也应该进行目视管理，结合颜色管理区分适当范围和极限范围。

姓名	插件1	插件2	插件3	自动插件	焊接	组装1	组装2	检测	包装
刘备	△	●	○	●	○	◎	◎	●	●
张飞	●	○	◎	○	◎			○	△
诸葛亮	○	△	●	△	◎			◎	
曹操	◎		○				○	●	
董卓		○		●		○			◎
貂蝉	◎		△		●				○
吕布	△		●	◎		●		○	◎
周瑜	●	○			△		●		

备注：△—计划学习；○—基本掌握；◎—完全掌握；●精通

上图为一公司员工技能评价表，这也是目视管理的一种形式，现场专门开辟目视管理看板，其上可以张贴此评价表。

◆检查表：它是 QC 七大手法中最简单也是使用得最多的手法，检查表使用简单、易于了解的标准化图形，人员只需填入规定的检查记号，再加以统计汇总其数据，即可提供量化分析或比对检查用，在 5S 活动中主要运用在检查环节。

★ 检查表的使用：

(1) 点检用，只记入好、不好的符号。

(2) 记录用，记录本评鉴的数据。检查表使用的时机是在推行 5S 运动告一段落后才能实施。

如此才能了解推行 5S 运动的成果，也会对 5S 运动产生信心。

办公场所诊断用5S检查表

项目	检 查 内 容	配分	得分	缺点事项
（一）整理	1. 是否定期实施红牌作战（清除不要品）？	4		
	2. 有无档案规定，并被清楚了解？	6		
	3. 桌子、文件架是否为必要最低限度？	4		
	4. 是否"没有必要的间隔"影响现场视野？	3		
	5. 桌子、文件架、通路是否有划分间隔？	3		
	小计	20		
（二）整顿	1. 建档规定是否确实被执行？	5		
	2. 文件等有无实施定位化（颜色、斜线）？	4		
	3. 磁碟片管理	4		
	4. 需要的文件、碟片能否马上取出？	5		
	5. 书柜、书架管理责任者？	3		
	6. 购置品有无规定放置处，并做补充规定？	4		
	小计	25		
（三）清扫	1. 地上、桌上是否杂乱？	3		
	2. 垃圾桶是否积得满满？	3		
	3. 管路、配线是否杂乱？	3		
	4. 供应开水处有无管理者标识？	3		
	5. 墙壁、玻璃是否保持干净	3		
	小计	15		
（四）清洁	1. OA机器有否保持干净？	3		
	2. 抽屉内是否杂乱？	3		
	3. 私有物品有无依规定放置？	3		
	4. 下班时桌上是否整齐？	3		
	5. 是否遵照穿着服装规定？	3		
	小计	15		
（五）教养	1. 是否有每周工作计划表来管理？	4		
	2. 部门的重点目标，目标管理是否被目视化？	4		
	3. 公告处有无规定，有无过期公告？	4		
	4. 接电话人不在，是否有留电话备忘录？	3		
	5. 是否活用目的表示板？	3		
	6. 有无文件分发及传阅规则？	4		
	7. 晨操是否积极参加？	3		
	小计	25		
合计		100		
评语				

机械加工、成型和安装工艺

<table>
<tr><td colspan="5" align="center">工厂现场诊断用5S检查表</td></tr>
<tr><th>项目</th><th>检查内容</th><th>配分</th><th>得分</th><th>缺点事项</th></tr>
<tr><td rowspan="6">（一）
整理</td><td>1. 是否定期实施红牌作战（清除不必要品）？</td><td>5</td><td></td><td></td></tr>
<tr><td>2. 有无不用或不急用的夹具、工具？</td><td>4</td><td></td><td></td></tr>
<tr><td>3. 有无剩料等近期不用的物品？</td><td>4</td><td></td><td></td></tr>
<tr><td>4. 有无"不必要的间隔"影响现场视野？</td><td>4</td><td></td><td></td></tr>
<tr><td>5. 作业场所是否有明确的区别标志？</td><td>3</td><td></td><td></td></tr>
<tr><td>小计</td><td>20</td><td></td><td></td></tr>
<tr><td rowspan="7">（二）
整顿</td><td>1. 仓库、储料室是否有规定？</td><td>4</td><td></td><td></td></tr>
<tr><td>2. 料架是否定位化？</td><td>4</td><td></td><td></td></tr>
<tr><td>3. 工具是否易于取用，不用找寻？</td><td>5</td><td></td><td></td></tr>
<tr><td>4. 工具是否有颜色区分？</td><td>4</td><td></td><td></td></tr>
<tr><td>5. 材料有无配置放置区，并加以管理？</td><td>5</td><td></td><td></td></tr>
<tr><td>6. 废弃品或不良品放置是否有规定，并加以管理？</td><td>3</td><td></td><td></td></tr>
<tr><td>小计</td><td>25</td><td></td><td></td></tr>
<tr><td rowspan="6">（三）
清扫</td><td>1. 作业现场是否杂乱？</td><td>3</td><td></td><td></td></tr>
<tr><td>2. 作业台是否杂乱？</td><td>3</td><td></td><td></td></tr>
<tr><td>3. 产品、设备有无脏污，附着灰尘？</td><td>3</td><td></td><td></td></tr>
<tr><td>4. 配置区划分线是否明确？</td><td>3</td><td></td><td></td></tr>
<tr><td>5. 作业段落或下班前有无清扫？</td><td>3</td><td></td><td></td></tr>
<tr><td>小计</td><td>15</td><td></td><td></td></tr>
<tr><td rowspan="6">（四）
清洁</td><td>1. 3S是否规划化？</td><td>5</td><td></td><td></td></tr>
<tr><td>2. 机械设备是否定期点检？</td><td>2</td><td></td><td></td></tr>
<tr><td>3. 是否遵照规定的服装穿着？</td><td>3</td><td></td><td></td></tr>
<tr><td>4. 工作场所有无放置私人物品？</td><td>3</td><td></td><td></td></tr>
<tr><td>5. 吸烟场所有无规定，并被遵守？</td><td>2</td><td></td><td></td></tr>
<tr><td>小计</td><td>15</td><td></td><td></td></tr>
<tr><td rowspan="7">（五）
教养</td><td>1. 有无日程管理表？</td><td>5</td><td></td><td></td></tr>
<tr><td>2. 需要的护具有无使用？</td><td>4</td><td></td><td></td></tr>
<tr><td>3. 有无遵照标准作业？</td><td>5</td><td></td><td></td></tr>
<tr><td>4. 有无异常发生时的对应措施？</td><td>3</td><td></td><td></td></tr>
<tr><td>5. 晨操是否积极参加？</td><td>3</td><td></td><td></td></tr>
<tr><td>6. 是否遵守开始、停止的规定？</td><td>5</td><td></td><td></td></tr>
<tr><td>小计</td><td>25</td><td></td><td></td></tr>
<tr><td colspan="2">合计</td><td>100</td><td></td><td></td></tr>
<tr><td colspan="2">评语</td><td></td><td></td><td></td></tr>
</table>

◆定点拍摄：在拍摄活动前后的情况时，在同一场所按同一方法进行拍摄，以容易比较变化前后的状况，让员工知道改善进度和改善效果。

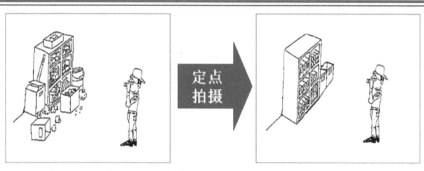

在同一场所按同一方法拍摄活动前后的现场实际情况

5S活动推行的组织管理

5S活动就像ISO认证活动一样，它的推行需要管理层的决心，所以建立一个公司内部的5S推行委员会可以大大提高全员参与的程度并保证检查监督工作的实施。

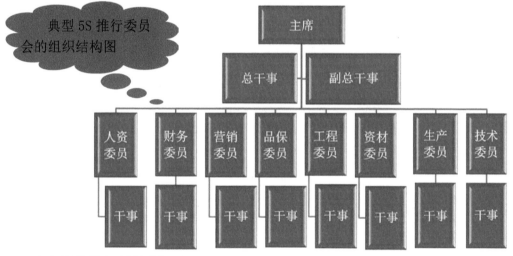

典型5S推行委员会的组织结构图

★各职位的主要职责如下：

1. **推行委员会**负责推行5S方针和目标的制定，确定实施方案。

2. **主席**负责委员会的运作、统筹、指挥和监督工作(一般由总经理或公司法人担任)。

3. **总干事**负责策划推行方案，协助主席处理委员会事务，主导中高层员工及5S干事教育训练(一般由副总经理担任)。

4. **副总干事**负责协助总干事，主导基层员工5S教育训练，召集会议筹备，组织检查，5S评比及评比结果的统计和公布(一般由品保部经理担任)。

5. **委员**负责参与制订5S活动方案，本部门5S活动监督及责任代表(一般由部门副职担任)。

6. **干事**负责协助参与制订5S活动方案，5S稽查评比，落实措施和反馈基层意见。

机械加工、成型和安装工艺

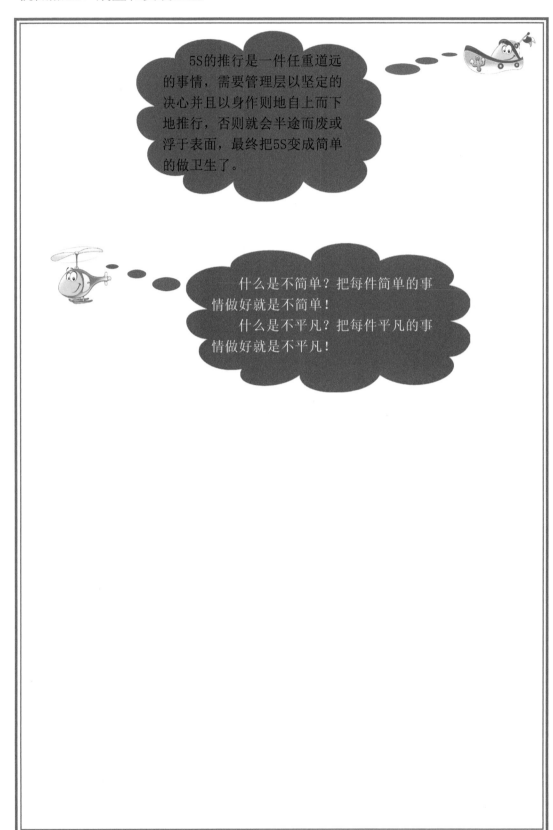

5S methodology

5S is a workplace organization method that uses a list of five Japanese words: *seiri* (整理), *seiton* (整頓), *seisō* (清扫), *seiketsu* (清洁), and *shitsuke* (素养). These have been translated as "Sort", "Set In order", "Shine", "Standardize" and "Sustain". The list describes how to organize a work space for efficiency and effectiveness by identifying and storing the items used, maintaining the area and items, and sustaining the new order. The decision-making process usually comes from a dialogue about standardization, which builds understanding among employees of how they should do the work.

In some quarters, 5S has become 6S, the sixth element being safety.

Other than a specific stand-alone methodology, 5S is frequently viewed as an element of a broader construct known as *visual control*, *visual workplace*, or *visual factory*. Under those (and similar) terminologies, Western companies were applying underlying concepts of 5S before publication, in English, of the formal 5S methodology. For example, a workplace-organization photo from Tennant Company (a Minneapolis-based manufacturer) quite similar to the one accompanying this article appeared in a manufacturing-management book in 1986.

The origins of 5S

5S was developed in Japan and was identified as one of the techniques that enabled Just in Time manufacturing.

Two major frameworks for understanding and applying 5S to business environments have arisen, one proposed by Osada, the other by Hirano. Hirano provided a structure to improve programs with a series of identifiable steps, each building on its predecessor. As noted by John Bicheno, Toyota's adoption of the Hirano approach was "4S", with Seiton and Seiso combined.

A precursor development to the Japanese system of management was outlined by Alexey Gastev's development and the Central Institute of Labour (CIT) in Moscow.

The 5S

There are five 5S phases. They can be translated from the Japanese as "sort", "set in order", "shine", "standardize", and "sustain". Other translations are possible.

Sort (Seiri)

Seiri is sorting through all items in a location and removing all unnecessary items from the location.

Goals:

◆ Reduce time loss looking for an item by reducing the number of items.

◆ Reduce the chance of distraction by unnecessary items.

◆ Simplify inspection.

◆ Increase the amount of available, useful space.

◆ Increase safety by eliminating obstacles.

Implementation:

◆ Check all items in a location and evaluate whether or not their presence at the location is useful or necessary.

◆ Remove unnecessary items as soon as possible. Place those that cannot be removed immediately in a "red tag area" so that they are easy to remove later on.

◆ Keep the working floor clear of materials except for those that are in use to production.

Set in order (Seiton)

Seiton is putting all necessary items in the optimal place for fulfilling their function in the workplace.

Goal:

Make the workflow smooth and easy.

Implementation:

◆ Arrange work stations in such a way that all tooling / equipment is in close proximity, in an easy to reach spot and in a logical order adapted to the work performed. Place components according to their uses, with the frequently used components being nearest to the workplace.

◆ Arrange all necessary items so that they can be easily selected for use. Make it easy to find and pick up necessary items.

◆ Assign fixed locations for items. Use clear labels, marks or hints so that items are easy to return to the correct location and so that it is easy to spot missing items.

Shine (Seiso)

Seiso is sweeping or cleaning and inspecting the workplace, tools and machinery on a regular basis.

Goals:

◆ Prevent deterioration.

◆ Keep the workplace safe and easy to work in.

◆ Keep the workplace clean and pleasing to work in.

◆ When in place, anyone not familiar with the environment must be able to detect any problems within 50 feet in 5 sec.

Implementation:

◆ Clean the workplace and equipment on a daily basis, or at another appropriate (high frequency) cleaning interval.

◆ Inspect the workplace and equipment while cleaning.

Standardize (Seiketsu)

Seiketsu is to standardize the processes used to sort, order and clean the workplace.

Goal:

Establish procedures and schedules to ensure the repetition of the first three "S" practices.

Implementation:

◆ Develop a work structure that will support the new practices and make it part of the daily routine.

◆ Ensure everyone knows their responsibilities of performing the sorting, organizing and cleaning.

◆ Use photos and visual controls to help keep everything as it should be.

◆ Review the status of 5S implementation regularly using audit checklists.

Sustain/Self-discipline (Shitsuke)

Shitsuke or sustain the developed processes by self-discipline of the workers. Also translates as "do without being told".

Goal:

Ensure that the 5S approach is followed.

Implementation:

◆ Organize training sessions.

◆ Perform regular audits to ensure that all defined standards are being implemented and followed.

◆ Implement improvements whenever possible. Worker inputs can be very valuable for identifying improvements.

◆ When issues arise, identify their cause and implement the changes necessary to avoid their recurrence.

参 考 文 献

[1] 冯学堂. 电火花加工[M]. 北京:中国劳动出版社,1997.

[2] 于文强,张丽萍. 机械制造基础[M]. 北京:清华大学出版社,2010.

[3] 王凤平. 机械制造工艺学[M]. 北京:机械工业出版社,2011.

[4] 郝兴明. 金属工艺学[M]. 北京:国防工业出版社,2012.